TECHNOLOGICAL TRAJECTORIES AND THE
HUMAN
ENVIRONMENT

Edited by

Jesse H. Ausubel
and
H. Dale Langford

NATIONAL ACADEMY OF ENGINEERING

NATIONAL ACADEMY PRESS
Washington, D.C. 1997

NATIONAL ACADEMY PRESS • 2101 Constitution Avenue, NW • Washington, DC 20418

The National Academy of Engineering was established in 1964 under the charter of the National Academy of Sciences as a parallel organization of outstanding engineers. It is autonomous in its administration and in the selection of its members, sharing with the National Academy of Sciences the responsibility for advising the federal government. The National Academy of Engineering also sponsors engineering programs aimed at meeting national needs, encourages education and research, and recognizes the superior achievements of engineers. Wm. A. Wulf is interim president of the National Academy of Engineering.

This set of papers first appeared in *Daedalus* (Vol. 125(3), 1996). This volume has been reviewed by a group other than the authors according to procedures approved by a National Academy of Engineering report review process. The interpretations and conclusions expressed in the papers are those of the authors and are not presented as the views of the council, officers, or staff of the National Academy of Engineering.

The activity that led to this publication was done in collaboration with the Electric Power Research Institute, The Rockefeller University, and the American Academy of Arts and Sciences. Funding for the publication was provided by the Andrew W. Mellon Foundation and the National Academy of Engineering Technology Agenda Program.

Library of Congress Cataloging-in-Publication Data

Technological trajectories and the human environment / edited by Jesse
    H. Ausubel and H. Dale Langford.
        p. cm.
    Includes bibliographical references and index.
    ISBN 0-309-05133-9 (alk. paper)
       1. Environmental management. 2. Technological innovations—
Environmental aspects. 3. Energy—Environmental aspects.
    4. Materials—Environmental aspects. I. Ausubel, Jesse.
    II. Langford, H. Dale.
    GE300.T43   1997
    363.7—dc21   96-48427
                                                 CIP

Cover art: *Figure in a Landscape*, oil on canvas, courtesy of the artist, Susan Bee, New York City.

This book is printed on recycled paper.

Printed in the United States of America

Preface

Among our largest and hardest problems are the unwanted environmental transformations associated with huge numbers of people and huge flows of material and energy. In 1988 the National Academy of Engineering launched an effort to increase understanding of long-term interactions between the environment and technological change and to identify opportunities to embed industry and its products more intelligently within nature. This effort is yielding a new discipline, industrial ecology, as well as the practical progress that comes from the stimulation of fresh networks of creative engineers and other experts.

Technological Trajectories and the Human Environment places in context the progress we are seeing. It also offers frameworks for understanding energy and materials that may help enterprises as well as society more broadly to select profitable and environmentally sound courses. And it draws our attention to some technologies of great promise. It does not set out a spectrum of possible futures, nor does it provide a complete picture by including, for example, military and medical technologies. Rather, this volume's authors provide scope and imagination in understanding technological evolution over the past century in energy, agriculture, and a few other fields of utmost importance to the environment. They speculate about what might be achieved if engineering and science continue to advance. In other words, suppose in these fields, societies continue to invent and innovate and diffuse technologies as has been the rule in the industrial era. Then what is the outlook for the environment? Several of the papers, however, also allude to the unintended, sometimes harmful consequences of technology, which are another lesson history teaches. Subsequent activities must probe more fully the possible unintended consequences of some of what the authors foresee, as well as the considerable obstacles to achieving the promise laid before us.

The volume complements earlier NAE books in this field published by the National Academy Press, including *Technology and Environment* (J. H. Ausubel and H.E. Sladovich, eds., 1989), *The Greening of Industrial Ecosystems* (B.R. Allenby and D.J. Richards, eds., 1994), and *Engineering within Ecological Constraints* (P. Schulze, ed., 1996), as well as the special section of Vol. 89(3)(1992) of the *Proceedings of the National Academy of Sciences of the USA* devoted to industrial ecology edited by NAE member Kumar Patel. Books soon to appear are *The Industrial Green Game: Implications for Environmental Design and Management*, which draws on international case studies; *The Ecology of Industry: Sectors and Linkages*, which looks at best practice in several industry sectors; and *Environmental Performance Measures and Ecosystem Condition*.

Many members of the NAE have contributed to our efforts in technology and environment, but I would especially like to thank Robert A. Frosch and Robert M. White for their sustained leadership, as well as Robert Herman and Chauncey Starr in the present volume. Indispensable intellectual as well as organizational assistance in this field has come from past NAE J. Herbert Hollomon Fellows Braden Allenby and Peter Schulze, as well as past and present NAE staff members Jesse Ausubel, Bruce Guile, and Deanna Richards. The continuing financial support of the Andrew W. Mellon Foundation for the Academy's work in industrial ecology is greatly appreciated.

The project that resulted in this volume benefited additionally from partnerships with the Electric Power Research Institute, the Program for the Human Environment of the Rockefeller University, and the American Academy of Arts and Sciences. The contents first appeared as "The Liberation of the Environment" in the summer 1996 issue of *Daedalus*, the journal of the American Academy, through the good auspices of its editor, Stephen Graubard. Former NAE editor H. Dale Langford teamed with Jesse Ausubel and Phyllis Bendell, the associate editor of *Daedalus*, to handle the innumerable steps between concept and publication. Advisers, participants in a preparatory workshop, and reviewers assisting with the volume (in addition to the authors) included David T. Allen, Brian J.L. Berry, Joshua Lederberg, Helga Nowotny, Stephen C. Peck, Vernon W. Ruttan, Andrew R. Solow, John H. Steele, Kurt E. Yeager, and Norton Zinder. I am confident readers will share my gratitude to all for opening wide our eyes to important possibilities for improving the ways we live and work.

Wm. A. Wulf
Interim President
National Academy of Engineering

Introduction:
Technological Trajectories and the Human Environment

ROBERT M. WHITE

The search for an economic system that can provide both more and better goods and services compatible with the long-run quality of the planetary environment has intensified with each recent decade. Industries and governments, as well as individuals, struggle to find orientation. In 1992 Rio de Janeiro hosted an unprecedented meeting of 130 heads of state to seek ways of harmonizing environment and development. But are environment and development truly the conflict?

At least two further questions appear to underlie this great question. Are the directions of invention and innovation, that is, the trajectories, in fact toward the lessening of pressures on the environment? And, if they are, are the rates of diffusion of innovations likely to be rapid enough to raise the quality of life for a large majority of people? To shed light on these questions, the National Academy of Engineering, Electric Power Research Institute (EPRI), and the Program for the Human Environment of the Rockefeller University collaborated in an effort to explore "technological trajectories and the human environment." The main fruit of that effort is this set of essays.

Collectively, the essays strike an optimistic note on a topic that generally evokes pessimism. The logic is sharp and the evidence surprisingly plentiful. Vast efficiencies can be achieved with respect to energy (Nakićenović), land (Waggoner), and materials (Wernick et al.). Indeed, we can begin to envision a quite different "industrial ecology" in which the notion of waste largely disappears (Frosch). Superb technical possibilities exist consistent with long-term historical developments and with one another (Ausubel and Marchetti, Starr).

v

And a new philosophical turn may relocate humans more constructively in nature (Meyer-Abich).

But we must also acknowledge cautions. The time period for a change can be considerable, 50 years or so for a major system (Grübler); appetites for goods and services appear unsated (Schipper); and solving problems in one time frame may create larger, harder ones over longer periods (Kates). Still, my overall sense is that the trajectories do point, in Ausubel's evocative phrase, toward a "liberation of the environment." Perhaps the Rio meeting really was a turning point.

The scope of this exploration is due in large part to Chauncey Starr, who insisted that we look many centuries back in order to locate ourselves in the present and to try to see decades and even centuries forward; and to Jesse Ausubel, who helped shape each essay. I hope this volume meets the wishes with which we began this effort: to provide stable, useful reference points for individuals as well as enterprises and governments on matters of the utmost long-run importance.

Contents

TECHNOLOGICAL TRAJECTORIES AND THE
HUMAN ENVIRONMENT

Technological Trajectories and the Human Environment.
Pp. 1–13. Washington, DC: National Academy Press

The Liberation of the Environment

JESSE H. AUSUBEL

The passage of time has connected the invention of the wheel with more than ten million miles of paved roads around the world today, the capture of fire with six billion tons of carbon going up in smoke annually. Must human ingenuity always slash and burn the environment? This essay and this volume suggest a more hopeful view. Indeed, the liberator of our title is human culture. Its most powerful tools are science and technology. These increasingly decouple our goods and services from demands on planetary resources.

Most observers emphatically designate the present as a period of intense environmental degradation. Surely, human numbers must weigh heavily, and they are highest now. Present world population stands at about 5.7 billion and each month increases by a number equivalent to the population of Sweden, Somalia, or New Jersey.

But for what period should we feel nostalgia? Has there been a golden age of the human environment? When was that age?

- In 1963, before the United States and Soviet Union signed the Limited Test Ban Treaty—after more than four hundred nuclear explosions in the atmosphere?

- In 1945, after much of the forest in Europe had been cut to provide fuel to survive World War II?

- In 1920, when coal provided three quarters of global energy, and choking smogs shrouded London and Pittsburgh?

1

- In 1870, when the Industrial Revolution boomed without filters in Silesia, Manchester, and Massachusetts?
- In 1859, before Edwin Drake first drew petroleum from an underground pool in Pennsylvania, when hunters slaughtered tens of thousands of whales for three million gallons of sperm oil to light American lamps?
- In the 1840s, when land-hungry farmers, spreading across North America, Australia, and Argentina, broke the plains and speedily shaved the native woods and grasses?
- In 1830, when cholera epidemics in many cities and towns literally decimated the populations that dumped their wastes in nearby waters?
- In 1700, when one hundred thousand mills interrupted the flow of every stream in France?
- In the late 1600s, when dense forests, once filled with a diversity of life, became seas of sugarcane in coastal Brazil and the Caribbean?
- In 1492, before Columbus stimulated reciprocal transatlantic invasions of flora and fauna? (The Old World had no maize, tomatoes, potatoes, green beans, groundnuts, sunflowers, cocoa, cotton, pineapple, vanilla, quinine, or rubber.)
- In the tenth century, before the invention of efficient chimneys, when people in cold climates centered their lives around a fireplace in the middle of a room with a roof louvered high to carry out the smoke—and much of the heat as well?
- In 55 B.C., when Julius Caesar invaded Britain and found less forest than exists today?
- In the centuries from Homer to Alexander, when the forests of the Eastern Mediterranean were cleared?
- Before the domestication of cows, sheep, pigs, and goats, when hunters caused a holocaust of wild creatures?
- In neolithic times, when building a house used up to thirteen tons of firewood to make the plaster for the walls and floor?

Environmental sins and suffering are not new (see, for example, Diamond, 1994; NACLA, 1991; Starbuck, 1964; and Turner et al., 1990). Humans have always exploited the territories within reach. The question is whether the technology that has extended our reach can now also liberate the environment from human impact—and perhaps even transform the environment for the better. My answer is that well-established trajectories, raising the efficiency with which people use energy, land, water, and materials, can cut pollution and leave much more soil unturned. What is more, present cultural conditions favor this movement.

ENERGY

Two central tendencies define the evolution of the energy system, as documented by Nebojša Nakićenović (this volume). One is that the energy system is freeing itself from carbon. The second is rising efficiency.

Carbon matters because it burns; combustion releases energy. But burnt carbon in local places can cause smog and in very large amounts can change the global climate. Raw carbon blackens miners' lungs and escapes from containers to form spills and slicks. Carbon enters the energy economy in the hydrocarbon fuels, coal, oil, and gas, as well as wood. In fact, the truly desirable element in these fuels for energy generation is not their carbon (C) but their hydrogen (H). Wood weighs in heavily at ten effective Cs for each H. Coal approaches parity with one or two Cs per H, while oil improves to two H per C, and a molecule of natural gas (methane) is a carbon-trim CH_4.

The historical record reveals that for two hundred years the world has progressively lightened its energy diet by favoring hydrogen atoms over carbon in our hydrocarbon stew. We can, in fact, measure this *decarbonization* in several different ways. As engineers, we can examine the changing ratio of the tons of carbon in the primary energy supply to the units of energy produced. From this perspective, the long-term, global rate of decarbonization is about 0.3 percent per year—gradual, to be sure, but enough to cut the ratio by 40 percent since 1860.

As economists, we can assess decarbonization as the diminishing requirement for carbon to produce a dollar's worth of economic output in a range of countries. Several factors dispose nations toward convergent, clean energy development. One is the changing composition of economic activity away from primary industry and manufacturing to services. End users in office buildings and homes do not want smoking coals. America has pared its carbon intensity of gross domestic product per capita per constant dollar from about three kilos in 1800 to about three-tenths of a kilo in 1990. The spectrum of national achievements also shows how far most of the world economy is from best practice. The present carbon intensity of the Chinese and Indian economies resembles those of America and Europe at the onset of industrialization in the nineteenth century.

Physical scientists can measure decarbonization in its elemental form, as the evolution of the atomic ratio of hydrogen to carbon in the world fuel mix. This analysis reveals the unrelenting though slow ascendance of hydrogen in the energy market. All the analyses imply that over the next century the human economy will squeeze most of the carbon out of its system and move, via natural gas, to a hydrogen economy (Ausubel, 1991). Hydrogen, fortunately, is the immaterial material. It can be manufactured from something abundant, namely water; it can substitute for most fuels; and its combustion to water vapor does not pollute.

Decarbonization began long before organized research and development in energy, and has continued with its growth. Many ways to continue along this trajectory have been documented. Still, the displacement of carbon remains the largest single environmental challenge facing the planet. Globally, people on average now use 1,000 kilograms of carbon per year compared, for example, to 120 kilograms of steel.

Part of economizing on carbon is economizing on energy more broadly. Efficiency has been gaining in the generation of energy, in its transmission and distribution, and in the innumerable devices that finally consume energy. In fact, the struggle to make the most of our fires dates back at least 750,000 years to the ancient hearths of the Escale cave near Marseilles. A good stove did not emerge until A.D. 1744. Benjamin Franklin's invention proved to be a momentous event for the forests and wood piles of America. The Franklin stove greatly reduced the amount of fuel required. Its widespread diffusion took a hundred years, however, because the colonials were poor, development of manufactures sluggish, and iron scarce (Reynolds and Pierson, 1942).

As Arnulf Grübler explains (this volume), we often fail to appreciate the speed and rhythms of social clocks. Many technological processes require decades or longer to unfold, in part because they cluster in mutually supportive ways that define technological eras every fifty years or so. The good news is that in a few decades most of our devices and practices will change, and major systems can become pervasive in fifty to one hundred years. It is also good news that latecomers to technological bandwagons can learn from the costly experiments of pioneers and that no society need be excluded from the learning. Evolutionary improvement and imitation transform the economy. Two percent per year may sound slow to a politician or entrepreneur, but maintained for a century it is revolutionary.

In energy and other sectors, the efficiency gains may have become more regular as the processes of social learning, embodied in science and technology, have taken root. In the United States since about 1800, the production of a good or service has required 1 percent less energy on average than it did the previous year. Nevertheless, embracing the full chain from the primary energy generator to the final user of light or heat, the ratio of theoretical minimum energy consumption to actual energy consumption for essentially the same mix of goods and services is still probably less than 5 percent (Ayres, 1989). No limit to increasing efficiency is near.

But engineers are working hard and getting results, as Ausubel and Marchetti dramatize (this volume) with a panorama of the past and future of electricity. In about 1700 the quest began to build efficient engines, at first with steam. Three hundred years have increased the efficiency of the devices from 1 to about 50 percent of their apparent limit. The technology of fuel cells may advance efficiency to 70 percent in another fifty years or so. While the struggle to improve generators spans centuries, lamps have brightened measurably with each decade. Edison's first lamp in 1879 offered about fifteen times the efficiency of a paraffin candle. The first fluorescent lamp in 1942 bettered Edison's by thirty times, and the gallium arsenide diode of the 1960s tripled the illumination efficiency of the fluorescent. Moreover, lamps are not the only means for illumination. The next century is likely to reveal quite new ways to see in the dark. For example, nightglasses, the mirror image of sunglasses, could make the night visible with a few milliwatts. We will speed efficiently to what we see. Using the same energy consumed by a present-day car, magnetically levitated trains in low-pressure tubes could carry a passenger several thousand kilometers per hour—connecting Boston and Washington in ten minutes.

LAND

Agriculture is by far the greatest transformer of the environment. Cities, paved roads, and the rest of the built environment cover less than 5 percent of the land in the forty-eight contiguous American states. Crops occupy about 20 percent of this land and pasture 25 percent. Crops cover 35 percent of France and 10 percent of China. Agriculture has consumed forests, drained wetlands, and voided habitats; the game is inherently to favor some plants and animals over others. Farms also feed us.

Yet since mid-century the amount of land used for agriculture globally has remained stable; and, as Paul Waggoner explains (this volume), the stage is set to reduce it. A shift away from eating meat to a vegetarian diet could roughly halve our need for land. More likely, diets will increase in meat and calories; under such conditions, the key will be the continuation of gains in yield resulting from a cluster of innovations including seeds, chemicals, and irrigation, joined through timely information flows and better-organized markets.

In fact, US wheat yields have tripled since 1940, and corn yields have quintupled. Despite these accomplishments, the potential to increase yields everywhere remains astonishing—even without invoking such new technologies as the genetic engineering of plants. The world on average grows only about half the corn per hectare of the average Iowa farmer, who in turn grows only about half the corn of the top Iowa farmer. Importantly, while all have risen steadily for decades, the production ratio of these performers has not changed much. Even in Iowa the average performer lags more than thirty years behind the state of the art. While cautious habits and other factors properly moderate the pace of diffusion of innovations, the effects still accumulate dramatically. By raising wheat yields fivefold during the past four decades, Indian farmers have in practice spared for other purposes an area of cropland roughly equal to the area of the state of California.

What is a reasonable outlook for the land used to grow crops for ten billion people, a probable world population sixty or seventy years hence? Future calories consumed per person will likely range between the 3,000 per day of an ample vegetarian diet and the 6,000 that includes meat. If farmers fail to raise global average yields, people will have to reduce their portions to keep cropland to its current extent. If the farmers can lift the global average yield about 1.5 percent per year over the next six or seven decades to the level of today's European wheat, ten billion people can enjoy a 6,000-calorie diet and still spare close to a quarter of the present 1.4 billion hectares of cropland. The quarter spared, fully

300 million hectares, would equal the area of India. Reaching the level of today's average US corn grower would spare for ten billion people half of today's cropland for nature, an area larger than the Amazon basin—even with the caloric intake of today's American as the diet.

The present realities of large amounts of land in Europe and North America reverting from farm to woodland, and high public subsidies to farmers, make the vision more immediate.[1] Beyond a world of ten billion people, it is not crazy to think of further decoupling food from land. For more green occupations, today's farmers might become tomorrow's park rangers and ecosystem guardians. In any case, the rising yields, spatial contraction of agriculture, and sparing of land are a powerful antidote to the current losses of biodiversity and related environmental ills.

WATER

Watts and hectares are yielding more. What about water? Chauncey Starr (this volume) points out that water is both our most valuable and most wasted resource. In the United States, total per capita water withdrawals quadrupled between 1900 and 1970. Consumptive use increased by one-third between just 1960 and the early 1970s, to about 450 gallons per day. However, since 1975, per capita water use has fallen appreciably, at an annual rate of 1.3 percent (US Geological Survey, 1993). Absolute US water withdrawals peaked about 1980.

Alert to technology as well as costs, industry leads the progress, though it consumes a small fraction of total water. Total industrial water withdrawals plateaued a decade earlier than total US withdrawals and have dropped by one-third, more steeply than the total. Notably, industrial withdrawals per unit of GNP have dropped steadily since 1940, from fourteen gallons per constant dollar to three gallons in 1990. Chemicals, paper, petroleum refining, steel, food processing, and other sectors have contributed to the steep dive (US Geological Survey, 1987). Not only intake but discharge per unit of production are perhaps one-fifth of what they were fifty years ago.

Law and economics as well as technology have favored frugal water use. Legislation, such as the US Clean Water Act of 1972, encouraged the reduction of discharges, recycling, and conservation, as well as shifts in relative prices. Better management of demand reduced water use in the Boston area from 320 million gallons per day in 1978 to 240 million gallons in 1992 (Stakhiv, 1996).

Despite such gains, the United States is a long way from exemplifying the most-efficient practice. Water withdrawals for all users in the industrialized countries span a tenfold range, with the United States and Canada at the highest end (OECD, 1991). Allowing for differences in major uses (irrigation, electrical cooling, industry, public water supply), large opportunities for reductions remain. In the late 1980s wastewaters still made up over 90 percent of measured US hazardous wastes. Importantly, as agriculture contracts spatially, its water demand will likewise tend to shrink.

In the long run, with much higher thermodynamic efficiency for all processes, removing impurities to recycle water will require small amounts of energy. Dialytic membranes open the way to such efficient purification systems. Because hydrogen will be, with electricity, the main energy carrier, its combustion may eventually provide another important source of water, perhaps 50 gallons per person per day at the level of final consumers, or about one-fourth the current withdrawal in water-prudent societies such as Denmark.

MATERIALS

We can reliably project decarbonization, food decoupled from acreage, and more efficient water use. What about an accompanying *dematerialization*? Wernick, Herman, Govind, and Ausubel define (this volume) dematerialization primarily as the decline over time in the weight of materials used to meet a given economic function. This dematerialization too would spare the environment. Lower materials intensity of the economy could translate into preservation of landscapes and natural resources, less garbage to sequester, and less human exposure to hazardous materials.

In fact, the intensity of use of diverse primary materials has plummeted over the twentieth century. Lumber, steel, lead, and copper have lost relative importance, while plastics and aluminum have expanded. Many products—for example, cars, computers, and beverage cans—have become lighter and often smaller. Although the soaring numbers of products and objects, accelerated by economic growth, raised municipal waste in the United States annually by about 1.6 percent per person in the last couple of decades, trash per unit of GDP dematerialized slightly.

The logic of dematerialization is sound. Over time new materials replace old, and theoretically each replacement should improve material properties per unit of quantity, thus lowering the intensity of use. Furthermore, as countries develop, the intensity of use of a given material (or system) declines as each country arrives at a similar level of development. The new arrivals take advantage of learning curves throughout the economy.

But superior materials also tend to make markets grow and thus take a kind of revenge on efficiency, offsetting the environmental benefits of each leaner, lighter object by enabling swarms of them to crowd our shelves. And our shelves lengthen. In Austin, Texas, the residential floor area available per person almost doubled in the past forty-five years—unsurprising when we consider that five people resided in the average US home in 1890 and 2.6 do now.

So far, trends of dematerialization are equivocal. Yet, as Robert Frosch theorizes (this volume), the potential surely exists to develop superior industrial ecosystems that reduce the intensity of materials use in the economy, minimize wastes, and use persisting wastes nutritiously in new industrial food webs. Since 1990 recycling has accounted for over half the metals consumed in the United States, up from less than 30 percent in the mid 1960s (see Wernick and Ausubel, 1995). The trick is to make waste minimization a property of the industrial system even when it is not completely a property of an individual process, plant, or industry. Advancing information networks may help by offering cheap ways to link otherwise unconnected buyers and sellers to create new markets or waste exchanges.

LIBERATION FROM THE ENVIRONMENT

I have focused primarily on trajectories, strategies, and technologies that lessen pollution and conserve landscape. It would hardly make sense to do so unless we wish to expand human notions of the rights of other species to prosper or at least compete. Klaus Michael Meyer-Abich explicitly argues (this volume; see also Meyer-Abich, 1993) that we must stand up for the "co-natural world," with which humans share Earth. We must take seriously the Copernican insight about Earth's position in the cosmos and not simply replace geocentricism with anthropocentricism. As advised by the great early nineteenth-century natural historian Alexander von Humboldt, we should participate in the whole as part of a part of a part of it, together with others. We may draw parallels between expanding notions of democracy and enfranchisement *within* human societies with respect to class, gender, and race, and our broadening view of the ethical standing of trees, owls, and mountains.

Yet the condition for our widespread willingness to take the Copernican turn is surely the successful protections we have achieved for our own health and safety. Recall how deaths from the human environment have changed during the last century or two (Ausubel et al., 1995; McKinlay and McKinlay, 1977).

First, consider "aquatic killers" such as typhoid and cholera, the work of bacteria that thrive in water polluted by sewers. In 1861 Queen Victoria's husband, Prince Albert, died of typhoid fever reportedly contracted from Windsor's water. Indeed, until well into the nineteenth century, townsfolk drew their water from ponds, streams, cisterns, and wells. They threw wastewater from cleaning, cooking, and washing on the ground, into a gutter, or into a cesspool lined with broken stones. Human wastes went into privy vaults—shallow holes lined with brick or stone, close to home, sometimes in the cellar. In 1829, New Yorkers deposited daily about one hundred tons of excrement into the city's soil.

Between 1850 and 1900 the share of the American population in towns grew from about 15 to about 40 percent. The number of cities with populations over fifty thousand grew from ten to more than fifty. Overflowing privies and cesspools filled alleys and yards with stagnant water and fecal wastes. The environment could not be more propitious and convenient for typhoid, cholera, and other water-borne diseases. They reaped 11 percent of all American corpses in 1900.

But by 1900, towns were also building systems to treat their water and sewage. Financing and constructing such facilities took several decades. By 1940 the combination of water filtration, chlorination, and sewage treatment stopped most of the aquatic killers in the United States. Refrigeration in homes, shops, trucks, and railroad boxcars took care of much of the rest. Chlorofluorocarbons (CFCs), the agents in today's thinning of the ozone layer, were introduced in the early 1930s as a safer and more effective substitute for ammonia in refrigerators; the ammonia devices tended to explode.

More killers have come by air, including tuberculosis (TB), diphtheria, influenza and pneumonia, measles, and whooping cough, as well as scarlet fever and other streptococcal diseases. In some years during the 1860s and 1870s, TB was responsible for 15 percent of all deaths in Massachusetts. Earlier in the nineteenth century, diphtheria epidemics accounted for 10 percent of all deaths in some regions of the United States. Influenza A is believed to have caused the Great Pandemic of 1918–1919, when flu claimed about a quarter of all corpses in the United States and probably more in Europe. (My own existence traces directly to this pandemic; my grandfather's first wife and my grandmother's first husband both died in the pandemic, leading to the union that produced my father.)

Collectively, the aerial killers accounted for almost 30 percent of all deaths in America in 1900. Their main allies were urban crowding and unfavorable living and working conditions. The aerial diseases began to weaken a decade later than the aquatics, and then weakened by a factor of seven over thirty years. Credit goes to improvements in the built environment: replacement of tenements and sweatshops with larger and better-ventilated homes and workplaces. Credit is also due to medical interventions. However, many of these, including vaccines and antibiotics, came well after the aerial invaders were already in retreat.

Formerly, most aerial attacks occurred in winter, when people crowded indoors; most aquatic kills occurred in summer, when organic material ferments speedily. Thus, mortality in cities such as Chicago used to peak in summer and winter. In America and other industrialized countries in temperate zones, the twentieth century has seen a dramatic flattening in the annual mortality curve as the human environment has come under control. In these countries, most of the faces of death are no longer seasonal.

Thus, when we speak of technological development and environmental change, it is well to remember first that our surroundings often were lethal. Where development has succeeded and peace holds, we have made the water fresher, the air cleaner, and our shelters more resistant to the violence of the elements. In the United States, perhaps 5 rather than 50 percent of deaths now owe to environmental hazards and factors, including environmentally-linked cancers. The largest global change is that humans—vulnerable, pathetic mammals when naked—have learned how to control their environment. Science and technology are our best strategies for control, and our success is why we now number nearly six billion.

But here is a catch for *homo faber,* the toolmaker. Our technology not only spares resources but also expands the human niche, within particular time frames. As Robert Kates explains (this volume), the intertwining of population, resources, and technology looks quite different depending on the time frame that one uses. From the greatest distance, human population appears to have surged three times. The first was associated with the invention of toolmaking itself, lasted about a million years, and saw human numbers rise to five million. The second surge swelled our population a hundredfold to about five hundred million over the next eight thousand years, following the domestication of plants and animals. Today we are midway into a third great population surge, which may level off at eleven billion or so three to four hundred years after the modern scientific and industrial revolution began.

But if one looks instead at the size of populations of regions over thousands of years, what goes up eventually comes down. In Egypt, Mesopotamia, the Central Basin of Mexico, and the Mayan lowlands, reconstructed population records show waves in which the population at least doubled over a previous base and then at least halved from that high point. Social learning works, but not forever. Societies flourish but they also forget and fail.

Shortening the time scale to recent centuries, we observe above all a systematic change in vital rates. Many countries have passed through the "demographic transition" from high death and birthrates to low death and birthrates. Technology certainly accounts for much of the increase in child survival and longevity, but no one can securely explain the changes in fertility, which ultimately determines the size of humanity. With respect to technology and fertility, the "pill" and its possible successors—while certainly more reliable—do not introduce an essential discontinuity in birth control. Many strategies against conception have always existed; parents have always essentially controlled family size. Though technology can ease implementation, population stabilization is a cultural choice (Marchetti et al., 1996). Fertility rates have been falling in most nations and are below levels needed to replace the current populations in Europe and Japan, which may implode. Perhaps the idea of the small family, which originated in France around the time of the Revolution, will become the norm after 250 years.

Still, recent population growth, which peaked globally at 2.1 percent per year around 1970, is unprecedented. The effect is that in the coming interval of a few decades human society will need to house, nurture, educate, and employ as many more people as already live on Earth. In the present era of lengthening lives and rising numbers, it appears, rather ironically, that our environmental achievement has been to liberate us from the environment.

In fact, high incomes, great longevity, and large population concentrations have been achieved in every class of environment on Earth. We manufacture computers in hot, dry Phoenix and cool, wet Portland. We perform heart surgery in humid Houston and snowy Cleveland. Year round we grow flowers in the Netherlands and vegetables in Belgium. The metro in Budapest runs regardless of the mud that slowed Hungarians for a thousand years. In Berlin and Bangkok we work in climate-controlled office buildings. We have insulated travel, communications, energy generation, food availability, and almost all major social functions from all but the most extreme environmental conditions of temperature and wind, light and dark, moisture, tides, and seasons.

The Japanese have even moved skiing and sand beaches indoors. In the world's largest indoor ski center, Ski-Dome near Tokyo, the slope extends 490 meters by 100 meters, with a thrilling drop of 80 meters that satisfies the standards of the International Ski Federation for parallel slalom competition. On the South Island of Kyushu, Ocean-Dome encloses 12,000 square meters of sandy beach and an ocean six times the size of an Olympic pool, filled with 13,500 tons of unsalted, chlorinated water kept at a warm 28°C. A wave machine produces surf up to three-and-a-half meters high, enough for professional surfing. Palm trees and shipwrecks provide the context.

In fact, careful records of human time budgets show that not only New Yorkers and Indians but also Californians, reputed nature enthusiasts, average only about one-and-a-half hours per day outside (Jenkins et al., 1992). Fewer than 5 percent of the population of industrialized nations work outdoors. In developing countries, the number is plummeting and should be below 20 percent globally by 2050. As Lee Schipper shows (this volume), life-styles revolve around the household. The achievement of ten thousand years of human history is that we have again become cave dwellers—with electronic gadgets.

THE LIBERATION OF THE ENVIRONMENT

For most of history thick forests and arid deserts, biting insects and snarling animals, ice, waves, and heat slowed or stopped humans. We built up our strength. We burned, cut, dammed, drained, channeled, trampled, paved, and killed. We secured food, water, energy, and shelter. We lost our fear of nature, especially in the aggressive West.

But we also secured a new insecurity. Although we have often cultivated the landscape with judgment and taste, we now recognize that we have transformed more than may be needed or prudent. Certainly, we would redo many episodes given the chance, particularly to protect precious habitats.

Some of our most arrogant behavior has been recent. Together the United States and the Soviet Union rocked Earth with close to two thousand nuclear blasts during the Cold War. The French, British, Chinese, and Indians also signaled their presence. The fifty-year bombing spree appears finally to be nearing an end.

Attitudes worldwide toward nature, and perhaps inseparably toward one another as humans, are changing. "Green" is the new religion. Jungles and forests, commonly domains of danger and depravity in popular children's stories until a decade or two ago, are now friendly and romantic. The Amazon has been transformed into a magical place, sanctified by the ghost of Chico Mendes, the Brazilian rubber tapper. Environmental shrines, such as the Great Sarcophagus at Chernobyl, begin to fill the landscape. The characterization of animals, from wolves to whales, has changed. Neither the brothers Grimm nor Jack London could publish today without an uproar about the inhumanity of their ideas toward nature—and I would add, with regard to gender and race as well.

Although long in preparation, great cultural changes can sweep over us in decades once under way. Moreover, standing against them is hopeless when they come. Magyar nobles vigorously opposed the spread of Protestantism and in 1523 declared it punishable by death and by the confiscation of property; despite all the edicts, Protestantism took firm hold in Hungary. In the nineteenth century in Europe and America a rising moral feeling made human beings an illegitimate form of property. Within about fifty years most countries abolished slavery. Many countries vocally rejected women's suffrage at the outset of the twentieth century. Now, politicians, though still mostly male, would not dream of mentioning the exclusion of women from full citizenship in most parts of the world.

The builders of the beautiful home of the US National Academy of Sciences in Washington, D.C., inscribed it with the epigraph, "To science, pilot of industry, conqueror of disease, multiplier of the harvest, explorer of the universe, revealer of nature's laws, eternal guide to truth." Finally, after a very long preparation, our science and technology are ready also to reconcile our economy and the environment, to effect the Copernican turn.[2] In fact, long before environmental policy became conscious of itself, the system had set decarbonization in motion. A highly efficient hydrogen economy, landless agriculture, industrial ecosystems in which waste virtually disappears: over the coming century these can enable large, prosperous human populations to co-exist with the whales and the lions and the eagles and all that underlie them—if we are mentally prepared, which I believe we are.

We have liberated ourselves from the environment. Now it is time to liberate the environment itself.

ACKNOWLEDGMENTS

I am grateful to Rudolf Czelnai, Cesare Marchetti, Perrin Meyer, and Iddo Wernick for assistance.

NOTES

1. For discussion of the re-creation of the "Buffalo Commons" in the US Great Plains, proposed by geographers Deborah and Frank Popper, see Matthews (1992). For a net estimate of changes in land use from growth of cities as well as changes in farming and forestry in the United States over the next century, see Waggoner et al. (1996).

2. For more information on required rates and amounts of change, see Ausubel (1996).

REFERENCES

Ausubel, J. H. 1991. Energy and environment: The light path. Energy Systems and Policy 15(3):181–188.

Ausubel, J. H. 1996. Can technology spare the Earth? American Scientist 84(2):166–178.

Ausubel, J. H., P. Meyer, and I. K. Wernick. 1995. Death and the Human Environment: America in the 20th Century. Working paper, Program for the Human Environment, The Rockefeller University, New York.

Ayres, R. U. 1989. Energy Inefficiency in the US Economy: A New Case for Conservation. RR-89-12. Laxenburg, Austria: International Institute for Applied Systems Analysis.

Diamond, J. M. 1994. Ecological collapses of ancient civilizations: The Golden Age that never was. Bulletin of the American Academy of Arts and Sciences XLVII(5):37-59.

Jenkins, P. L., T. J. Phillips, E. J. Mulberg, and S. P. Hui. 1992. Activity patterns of Californians: Use of and proximity to indoor pollutant sources. Atmospheric Environment 26A(12):2141–2148.

Marchetti, C., P. Meyer, and J. H. Ausubel. 1996. Human population dynamics revisited with a logistic model: How much can be modeled and predicted? Technological Forecasting and Social Change 52:1–30.

Matthews, A. 1992. Where the Buffalo Roam. New York: Grove Weidenfeld.

McKinlay, J. B., and S. M. McKinlay. 1977. The questionable contribution of medical measures to the decline of mortality in the United States in the twentieth century. Milbank Quarterly on Health and Society (Summer):405–428.

Meyer-Abich, K. M. 1993. Revolution for Nature: From the Environment to the Co-Natural World. Cambridge, England, and Denton, Tex.: White Horse and University of North Texas Press.

NACLA (North American Congress on Latin America). 1991. The conquest of nature, 1492–1992. Report on the Americas 25(2).

OECD (Organization for Economic Cooperation and Development). 1991. The State of the Environment. Paris: OECD.

Reynolds, R. V., and A. H. Pierson. 1942. Fuel Wood Used in the United States: 1630–1930. Circular 641. Washington, D.C.: US Department of Agriculture.

Stakhiv, E. Z. 1996. Managing water resources for climate change adaptation. Pp. 243–264 in Adapting to Climate Change: Assessment and Issues, J. B. Smith, N. Bhatti, G. Menzhulin, R. Benioff, M. I. Budyko, M. Campos, B. Jallow, and F. Rijsberman, eds. New York: Springer-Verlag.

Starbuck, A. 1964. History of the American Whale Fishery from its Earliest Inception to 1876. Vol. 1. New York: Argosy-Antiquarian.

Turner, B. L. II, W. C. Clark, R. W. Kates, J. F. Richards, J. T. Mathews, and W. B. Meyer. 1990. The Earth as Transformed by Human Action. New York: Cambridge University Press.

US Geological Survey. 1987. Pp. 81–92 in National Water Summary 1987—Hydrologic Events and Water Supply and Use. Water Supply Paper 2350. Washington, D.C.: US Government Printing Office.

US Geological Survey. 1993. Estimated Use of Water in the United States in 1990. Circular 1081. Washington, D.C.: US Government Printing Office.

Waggoner, P. E., J. H. Ausubel, and I. K. Wernick. 1996. Lightening the tread of population on the land: American examples. Population and Development Review 22(3):531–545.

Wernick, I. K., and J. H. Ausubel. 1995. National materials metrics for industrial ecology. Resources Policy 21(3):189–198.

Technological Trajectories and the Human Environment. 1997.
Pp. 14–32. Washington, DC: National Academy Press.

Time for a Change:
On the Patterns of Diffusion of Innovation

ARNULF GRÜBLER

A MEDIEVAL PRELUDE

The subject of this essay is the temporal patterns of the diffusion of techno-logical innovations and what these patterns may imply for the future of the human environment.[1] But first let us set the clock back nearly one thousand years: return for a moment to monastic life in eleventh-century Burgundy.

Movement for the reform of the Benedictine rule led St. Robert to found the abbey of Cîteaux (Cistercium) in 1098. Cîteaux would become the mother house of some 740 Cistercian monasteries. About 80 percent of these were founded in the first one hundred years of the Cistercian movement; nearly half of the foundings occurred in the years between 1125 and 1155 (see Figure 1). Many traced their roots to the Clairvaux abbey founded as an offshoot of Cîteaux in 1115 by the tireless St. Bernard, known as the Mellifluous Doctor. The nonlinear, S-shaped time path of the initial spread of Cistercian rule resembles the diffusion patterns we will observe for technologies. The patterns of temporal diffusion do not vary across centuries, cultures, and artifacts: slow growth at the beginning, followed by accelerating and then decelerating growth, culminating in saturation or a full niche. Sometimes a symmetrical decline follows or a new growth pulse.

Over time the Cistercians also diffused in space. Their pattern of settlements shows significant differences in spatial density. The innovation origin, Burgundy, was home to the four major mother houses and hosted the highest spatial concen-tration of settlements. From there, daughter houses were founded ("regional subinnovation centers," in the terminology of spatial diffusion), from which

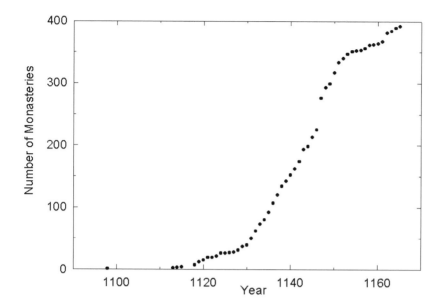

FIGURE 1   The initial diffusion of Cistercian monasteries in Europe. DATA SOURCE: Janauschek (1877).

Cistercians spread further into their respective hinterlands ("the neighborhood effect") and to other subregional centers, originating yet further settlements. The density of settlements decreased at the periphery, away from innovation centers, implying persistent regional diversity and disparities. The Cistercians also differentiated into "subfamilies," named after their respective parental houses. In fact, each subfamily followed its own pattern of settlements, regional specialization, and implementation of the Cistercian rule.

Some of the additions to the Cistercian rule were not genuine new settlements but "takeovers." For example, the existing Benedictine monastery of Savigny, with all its daughter houses, submitted to the rule of the Clairvaux Cistercians in 1147 and in turn became the mother house of all Cistercian settlements in the British Isles.

Despite distance and differentiation, all the monasteries communicated closely. The industrious Cistercians thus introduced and channeled influential innovations, including new agricultural practices and the water mill, throughout Europe in the thirteenth and fourteenth centuries. The British monks excelled in wool production. In fact, according to the Cistercian rule, settlements were to be located in remote, undeveloped areas. Thus, Cistercian monasteries became important local nodes for the colonization of land within Europe and, hence, for deforestation.

The Cistercian topology reveals a hierarchy of centers of creation and structured lines of spread. The patterns bear witness to the existence of *networks*. As we shall see, social and spatial networks, and their interactions, support and shape the diffusion process.[2]

INVENTION, INNOVATION, THEN DIFFUSION

In discussing the time for a change associated with a technology, it is necessary to consider invention and innovation as well as diffusion. Discourse now customarily distinguishes among these three concepts following the classic analyses made in the 1930s by the Austrian economist Joseph Schumpeter (1939). Invention is the first demonstration of the principal feasibility of a proposed new artifact or solution. Fermi's Chicago reactor demonstrated the feasibility of a controlled nuclear fission reaction (invention). In 1958, sixteen years after the inauguration of Fermi's pile, the Shippingport, Pennsylvania, reactor went into operation to generate commercial electric power (innovation). Some forty years later more than one hundred nuclear reactors now generate some 20 percent of the electricity in the United States (diffusion). Analogously, we might say St. Robert invented the Cistercian rule, St. Bernard innovated, and diffusion followed.

In fact, considering the Cistercian rule as a technology makes an important point. In the narrowest definition, technology is represented by the objects people make, axes and arrowheads and their updated equivalents. Anthropologists call them "artifacts"; engineers call them "hardware." But technology does not end here. Artifacts must be produced, that is, invented, designed, and manufactured. This process requires a larger system of hardware (machinery, a manufacturing plant), factor inputs (labor, energy, raw materials), and finally "software" (human knowledge and skills).

The third of these elements, which French scholars call *technique,* represents the disembodied aspect of technology, its knowledge base. Technique is required not only for the production of given artifacts but ultimately also for their *use,* both at the level of the individual and at the level of society. An individual must know, for example, how to drive a car; a society must know how to conduct an election. Organizational and institutional forms (including markets), social norms, and attitudes all shape how particular systems of production and use of artifacts emerge and function. They are the originating and selection mechanisms of particular artifacts (or combinations thereof) and set the rate at which they become incorporated into a given socioeconomic setting. This process of filtering, tailoring, and acceptance is technology diffusion.

Before discussing diffusion further, let us return to the prior processes, invention and innovation. In truth, a realistic history of social and technological innovations would consist mostly of nonstarters. The overwhelming share of inventions are ignored. And an analysis of several hundred major innovations over the past two centuries shows a typical span of about fifteen to forty years between invention and innovation (Mensch, 1975). Moreover, the existence of one or more possible innovations in itself hardly guarantees subsequent diffusion.

To appreciate the uncertainty in the early phases of technology development, let us look at a historical problem of technological hazard and environmental pollution from steam railways. In the early days of railroad expansion in the United States, sparks in the smoke from wood-burning steam locomotives caused a considerable fire hazard to both human settlements and forests (Basalla, 1988). Inventors and entrepreneurs registered more than one thousand patents on "smoke-spark arresters" during the nineteenth century in a futile search for a solution, which arrived finally not by an add-on technology but by the replacement of steam by diesel and electric locomotives. This large number of alternatives illustrates that diversity and experimentation are precursors to diffusion. Many are called, but few are chosen.

Moreover, what is chosen for diffusion is not necessarily the best. The selection of a particular technological alternative may not conform to *ex ante* or *ex post* judgments about optimality. Sometimes selection of a particular alternative stems from an accumulation of small, even random events, eventually "locking in" a particular configuration. Thereafter, positive feedback mechanisms yield increasing returns to adoption of the standardized alternative. We suspect that the standard gauge of railroads or the disk operating systems in use now in personal computers are not the "best" but simply prevailed at a certain time in history and therefore can only be dislodged with great difficulty (see Arthur, 1988).

What are the factors in setting the diffusion clock? One is simply opposition to change. Opposition to proposed and diffusing technologies always recurs. The most cited case is the Luddites, who destroyed knitting and other textile machinery between 1811 and 1816. A similar movement, led by Captain Swing, resisted the introduction of mechanical threshing in rural England in the 1830s. As shown in Figure 2, the opposition to the machines was itself an orderly diffusive process. The time it took for the craze to smash machines to spread—two weeks—shows that social interaction and communication were highly effective far in advance of modern transport and telephony. Although opposition causes uncertainty about the eventual fate of an innovation, it fulfills two important evolutionary roles. First, it can operate as a selection mechanism for rejecting socially unsustainable solutions or technologies. Second, it helps qualify technologies to respond to societal concerns, improving their performance and thus enabling further, even pervasive, diffusion.

In a classic article, Earl Pemberton (1936) provided many illuminating examples of curves of gradual cultural diffusion. The first country to introduce postage stamps was England in 1840. Such a good idea; yet it took close to fifty years for a sampling of thirty-seven independent states in Europe, North America, and South America to imitate. A more delicate idea, touching on the nature and

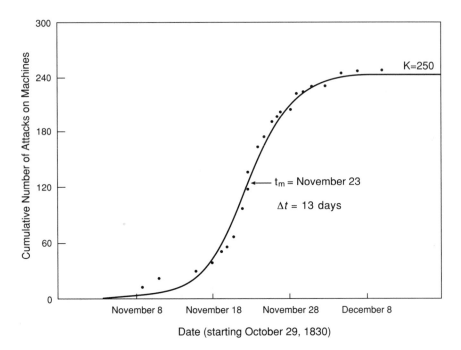

FIGURE 2 Resistance to technology as a diffusion process: number of threshing machines attacked during the Captain Swing Movement in England in 1830. NOTE: Actual data and a fitted three-parameter logistic curve. See endnote 5. DATA SOURCE: Hobsbawm and Rudé (1968).

control of the family, is the first compulsory school attendance law, enacted at the state level in the United States in 1847. It took fully eighty years, until 1927, for the last state then belonging to the United States to adopt similar legislation. These examples already emphasize that changes in technologies and social techniques are not one-time, discrete events but rather a process characterized by time lags and often lengthy periods of diffusion.

They also suggest that when diffusion succeeds, the forces and factors determining its speed and extent may change over time.[3] Performance, cost, fashion, and familiarity are among the considerations. Nevertheless, the diversity and complex interactions at the micro level appear often to lead to smooth, orderly behavior at the macro level, whether of Cistercians and Luddites, or, as we shall see, canals and passenger cars. Some theorists argue that orderly macroeconomic evolution requires such microeconomic diversity, which at first glance might instead seem likely to dissipate order (see Dosi et al., 1986; Silverberg, 1991; and Silverberg et al., 1988).

In addition to sociological and economic factors, straightforward, generic considerations appear to influence the speed of diffusion. The scope of technical change itself is a powerful one. We might distinguish four levels: 1) incremental improvements; 2) radical changes in individual technologies and artifacts; 3) changes in technology systems, that is, combinations of radical changes in technologies combined with organizational and managerial changes; and 4) changes in clusters and families of technologies and in associated organizational and institutional settings.[4] The latter levels of change, as well as larger system sizes, will likely entail longer times for diffusion (Grübler, 1991).

In sum, inventive and innovative activities provide the *potentials* for change. However, *diffusion* translates these potentials into changes in social practice. One abbey could not transform European agriculture; 740 did. Diffusive, largely imitative or repetitive phenomena are at the heart of the changes in society and its material structures, infrastructures, and artifacts. Thus, in the subsequent discussion, the analysis of time required for diffusion provides the central metric to analyze processes of social and technological change. Let us now try to grasp the main patterns.

THE DURATION OF DIFFUSION

We will consider an increasingly complex series of cases of technology diffusion, characterized by the environment in which diffusion processes operate. In the simplest case, an idea, practice, or artifact represents so radical a departure from existing solutions that it largely creates its own market niche. In practice, preexisting means for meeting basic social functions, such as transport and communication, are always present; nothing is truly new or free of competitors. Physicist Elliott Montroll (1978) called evolution a sequence of replacements. But clearly, some technologies enter much more accommodating environments than others.

The development of canals in the early nineteenth century offers a reasonable case of simple diffusion. In fact, the actual data on the growth of the canal network in the United States are approximated very well by a symmetrical growth curve, a three-parameter logistic equation in this case (Figure 3).[5] The estimated upper limit of the diffusion process, some 4,000 miles of canals, matches the historical maximum of 4,053 miles of canal in operation in 1851. The characteristic duration of diffusion (or Δt), defined as the time required for the process to unfold from 10 percent to 90 percent of its extent, is thirty-one years. The canals spread through the United States at about the same rate as the Cistercians initially spread through Europe. The entire canal diffusion cycle from 1 percent to 99 percent spans some sixty years. The year of maximum growth, or midpoint (t_m), occurred in 1835.

Subsequent major transport infrastructures, rails and roads, evolved along a dynamic pattern similar to canals, as Figure 4 illustrates (Grübler and Nakićenović, 1991). In the figure the sizes of individual networks have been

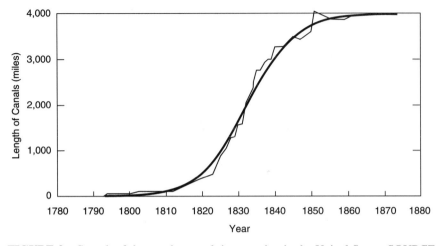

FIGURE 3   Growth of the canal network in operation in the United States. SOURCE: Grübler (1990).

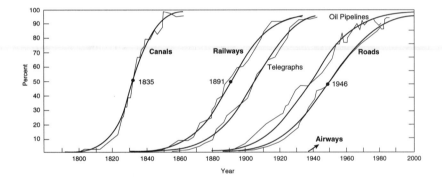

FIGURE 4   Growth of infrastructures in the United States as a percentage of their maximum network size. SOURCE: Grübler and Nakićenović (1991).

normalized for better comparability; in absolute extension, railways and surfaced road networks were one and two orders of magnitude larger, respectively, than canals at their maximum network length. Not surprisingly, the duration of the growth of railway and surfaced road networks is somewhat slower, Δt's of fifty-five and sixty-four years, respectively. Interestingly, we see the three major historic transport infrastructures spaced rhythmically apart in their development by a half century or so.

Transport infrastructures strongly influence nearly every aspect of daily life.[6]

Here we will comment only on their close relationship with other infrastructures. As Figure 4 suggests, the railway and the telegraph evolved together, as did the road network and the oil pipelines delivering the fuel for the cars on the roads. This synchronization illustrates technological interdependence and cross enhancement. Particular technologies and techniques do not diffuse in isolation but in a larger context, as we shall discuss below.

In fact, a new solution does not evolve in a vacuum but interacts with existing practices and technologies. One technology replaces or substitutes for another, with varying degrees of direct one-to-one competition. For example, after reaching its maximum size, the canal network declined rapidly because of vicious competition from railways. Looking at relative "market shares" of competing alternatives rather than at absolute volumes makes the interaction visible.

Probably the most famous case of technological substitution is motor cars for horses. In this case, the diffusion of one technological artifact, the passenger car, began simply by replacing another, the riding horse and the carriage. Looking at the absolute numbers of draft animals and cars in the United States (Figure 5), we see that the millions of horses and mules used for transport practically disappeared from the roads within fewer than three decades. Measured by a curve fit to a model of logistic substitution (for the model see Marchetti and Nakićenović, 1979), the duration of the replacement process (Δt) was only twelve years, fast enough to traumatize the oat growers and the blacksmiths (see Nakićenović,

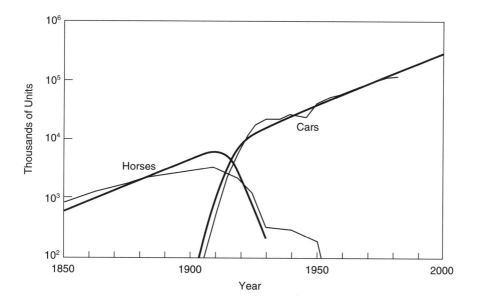

FIGURE 5 Number of nonfarm draft animals and automobiles. SOURCE: Nakićenović (1986).

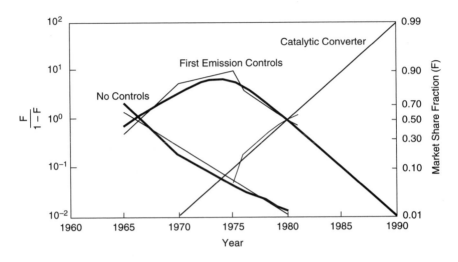

FIGURE 6   Diffusion of cars with first emission controls and catalytic converters in the United States, in fractional shares of total car fleet.   SOURCE: Nakićenović (1986).

1986). Interestingly, the diffusion of a modern anti-pollution device, the catalytic converter, also occurred with a Δt of twelve years in the United States (Figure 6). The reason is probably that the lifetime of the road vehicle has not changed since the horse-and-carriage era; the working lives of horses and cars both last about ten to twelve years.

The continuing growth of the car population in Figure 5 illustrates another dynamic feature of technological evolution: growth beyond the initial substitution or field of application. Use of the car grew initially by replacing horses. After completion of that process in the 1930s, new markets were created. Higher average speeds, greater reliability in all weather conditions, and other features opened chances both for competition with trains for long-distance travel and for short-distance commuting that created suburbs, which in turn created more demand for cars. Currently some 150 million passenger cars are registered in the United States, about 0.6 cars per capita.

Mention of the sequence of horses, trains, and cars brings us to consider the most realistic process of technological change: multiple competing technologies. In steel manufacturing as many as four technologies have competed simultaneously with decreasing and increasing market shares (Figure 7). The diffusion trajectories of the processes are diverse, with Δt's ranging from less than two decades (replacement of the crucible process) to nearly seven decades (diffusion of electric arc steel). These changes in process technology not only enabled significant expansion of production but mattered greatly from an environmental perspective. They coincided with changes in energy supplies toward higher qual-

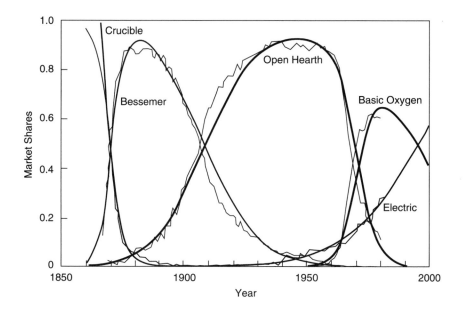

FIGURE 7   Process technology change in US steel manufacturing, in fractional shares of raw steel tonnage produced. SOURCE: Nakićenović (1987).

ity and cleaner energy carriers, consistent with the overall evolution of energy supply (see Nakićenović, this volume). Between 1800 and 1930 in the United States, one hundred million cords of hardwood are estimated to have been cut for charcoal for smelting iron (Reynolds and Pierson, 1942).

Let us now bring space back into our time picture. We have drawn examples so far from the United States. We commented at the outset about the patterns in space as well as the time of the diffusion of the Cistercian rule. Does the same hold true for a modern technology such as the motor car? Like Burgundy and its Cistercians, the United States was the earliest adopter of the car and has achieved the highest density of cars. Having started to adopt cars rapidly about the year 1910, America now has almost six hundred cars per thousand people. Having started in 1930, the United Kingdom now parks about four hundred cars per thousand people, while Japan parks about three hundred per thousand, having started the adoption process only in the 1950s. As Figure 8 suggests, empirical data from numerous countries show that later adopters manifest both an accelerated diffusion rate (shorter diffusion time) and a declining density of adoption as a function of the introductory date. The case of cars is corroborated by analysis of the declining adoption densities of "late-starters" in the railway development of the nineteenth century (Grübler, 1990).

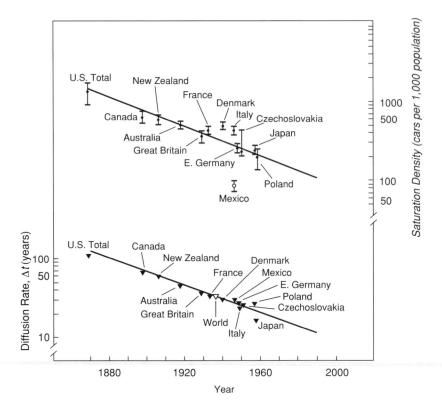

FIGURE 8   Passenger car diffusion at the global level: Catch-up, but at lower adoption levels. NOTE: Estimated saturation density and diffusion rates expressed as a function of the introduction date of the automobile. SOURCE: Grübler (1990).

The spread of railway networks in fact clearly shows how both spatial densities and the temporal rates of the adoption of technologies remain diverse. In the United States, the early innovation centers for railways on the East Coast and around the Great Lakes achieved by far the greatest spatial density of networks. Railway construction reached the West Coast some fifty years after the East Coast, and network densities remained significantly lower. In Europe, rails spread from the north of England in the 1820s to the rest of England and also to Belgium. By 1836 independent innovation centers had arisen in the Lyons region of France and Austria-Bohemia. The railway innovation wave spread from the early continental centers to cover most of Western and Central Europe by the 1850s. By the mid 1870s all of Eastern Europe, as well as most of European Russia, southern Scandinavia, and part of the Balkans, were networked. The final European subinnovation center was Greece, toward 1900. Rails penetrated the Albanian region almost a century after England. Starting first, England built a network (with attendant costs and benefits) one-third denser than Germany, almost twice the density of France, and ten or more times denser than other countries that might have appeared comparable at the outset of the railroad era.

In this light, we can ask, is the United States a likely guide for future mass-motorization globally? According to our understanding, no. Instead, the high density of cars in the United States results from specific initial conditions, including high individual mobility before the advent of the automobile and a long period of diffusion, which created precisely the conditions in life-style, spatial division of labor, and settlement patterns of an "automobile society." As Figure 8 indicates, heterogeneity in rates of diffusion and thus levels of adoption follows orders and thus is likely to persist, not only for railways and autos but in general for systems that diffuse globally. This perspective leads to lower-than-usual estimates of future demand for transport energy for China, for example (Grübler, 1992).

SEASONS OF SATURATION

We have noted that clusters of radical innovations and technology systems, interdependent and mutually cross-enhancing, give rise to families of technological innovations with associated new institutional and organizational settings. For example, the development of the automotive industry was contingent on developments in materials (high-quality steel sheets), the chemical industries (oil refining, in particular catalytic cracking), production and supply infrastructures (exploration and oil production, pipelines and gasoline stations), development of public infrastructures (roads), and a host of other technological innovations. The growth of the industry was based on a new production organization (Fordist mass production combined with Taylorist scientific management principles), yielding significant real-term cost reductions that made the car affordable to more social strata, thus changing settlement patterns, consumption habits of the population, and leisure activities. In turn, the automobile is just one artifact among many consumer durables now standard in every household in industrialized countries. These linkages multiply the effects of such techno-institutional clusters on the economy and society and account for their pervasive impact.

To quantify the emergence of technology clusters, I analyzed the history of a large sample of technologies for the United States (Grübler, 1990, 1991) Consistent with the definition of technology adopted here, the sample used in the analysis was not taken from the hard technology field alone. The cases included diffusion of energy, transport, manufacturing, agriculture, consumer durables, communication, and military technologies, as well as diffusion of economic and social processes, such as literacy, reduction of infant mortality, and changes in job classes. Two samples were analyzed. The first consisted of 117 diffusion cases that my colleagues at the International Institute for Applied Systems Analysis and I had studied ourselves (see Grübler, 1990; Marchetti, 1980; Marchetti and Nakićenović, 1979; and Nakićenović, 1986). The second sample was aug-

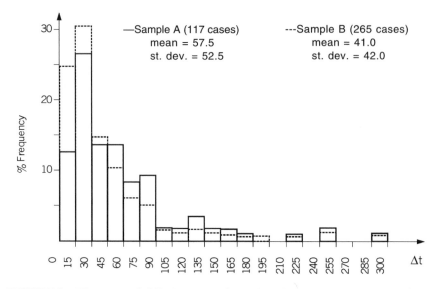

FIGURE 9   Histogram of diffusion rates of samples of 117 and 265 processes of technological, economic, and social change in the United States. NOTE: The Δt equals the time in years for a process to extend from 10 to 90 percent of its duration. St. dev. = standard deviation. SOURCE: Grübler (1990, 1991).

mented by additional, well-documented cases with a quantification of diffusion parameters that we found in the literature. This sample totaled 265 cases of innovation.

The profile of the diffusion rates, or Δt's, was quite similar for the two samples. The rates ranged from very short-term processes of only a few years to processes that extended over two to three centuries. The mean value ranged between forty and sixty years, with a standard deviation of about equal size (Figure 9). The largest number of diffusion processes in our samples have characteristic durations, Δt's, of between fifteen and thirty years.[7] If our diffusion studies had documented more of the seemingly numerous short-term phenomena such as clothing fashions, the profile of the histogram in Figure 9 would likely approach a "rank-size" or Zipf distribution in which the frequency of diffusion rates would be highest for fast processes and decline as the rates became slower.[8]

The good news for the human environment from our analysis is that the majority of artifacts and practices can be replaced within a few decades. However, some key processes have demonstrably long durations. For example, the global quests for improvements in the thermodynamic efficiency of prime movers and for the decarbonization of the energy system both clock in at about three hundred years (see Ausubel and Marchetti, this volume; Nakićenović, this volume). In general, pervasive transformations take time. The transformation of the US population from a society of farmers to manufacturers to service workers took some two hundred years (Herman and Montroll, 1972). Societies starting the move from brown to blue and to white collars later may accordingly move faster, but such all-embracing processes will never collapse to weeks and months.

We might summarize by saying that at any time, change in a society can be decomposed into a large number of diffusion (or substitution) processes with great variety in their rates. We can then ask whether aggregate measures exist for the average diffusion rate over time for the whole socioeconomic system and whether it changes. For such a measure, I calculated the average diffusion rates of the innovation samples, that is, the sum of the first derivatives of the diffusion (or substitution) trajectories at each point in time divided by the number of diffusion processes then occurring. This indicator is the diffusion equivalent of the annual GNP growth rate. The resulting measure rates the average annual technical (and economic and social) change at the country level (Grübler, 1990, 1991).

For the United States since 1800, the calculated average diffusion rate portrays clear peaks and troughs, which vary by a factor of two or more. The process of change is not gradual and linear but is instead characterized by long swings and discontinuities. In addition, rates of change tend to increase over time. This rise may reflect that the closer we approach the present, the more processes are included in the sample. However, the rising average rate of change could also result from the cumulative nature of technological change. Even though no individual diffusion process may proceed faster when compared to the past, the number and variety of artifacts (particularly those with faster turnover rates) are in fact much larger today than earlier. This could increase the average rate of change. In other words, while no individual technology or artifact diffuses faster than it did in the past (other things being equal), many more technologies and objects are in use, and thus more change. In any case, the analyses show pronounced discontinuities and also a decline in the diffusion rate in the decades after 1970, indicating an increase in saturation phenomena in the United States since then.

The fluctuations and discontinuities in the long-term rate of sociotechnical change result from the complex dynamics of the discontinuous rates at which individual innovations appear and from the different rates of absorption of these innovations in the socioeconomic system. Periods of accelerating rates appear to indicate the emergence of a technology cluster in which a large number of interrelated innovations diffuse into the economic and social environment. These in turn contribute, by means of backward and forward linkages, to prolonged periods of economic growth.

Periods in which progressively more and more innovations enter their saturation phase of diffusion follow the growth periods. Thus, each major peak in the average rate of change characterizes the start of saturation of a corresponding cluster or family of diffusion processes. This "season of saturations" results in a significant decline in the average rate of technical and social change and, through market saturation and a decrease in investments, also contributes to a slowdown in economic growth.

Presumably many inventions of the past few decades now await their chance to become successful innovations. Were they included, these could reverse the recent downward trend in the rate-of-change curve by the late 1990s. Then the successful innovations, after a slow initial diffusion, would enter into the rapid, indeed exponentially growing part of their life cycle.

The turning points in the rates of diffusion of technological and social innovations coincide with the turning points of so-called long-waves of economic growth as identified by several researchers (Marchetti, 1980; van Duijn, 1983; Vasko, 1987). In the analysis of US data, the peaks—the maxima in the rate of sociotechnical change and the onset of leveling off and saturation phenomena—occurred in 1840, 1912, and 1970, respectively. Troughs, maxima of saturation periods and the slow beginning of a new phase of accelerated sociotechnical change, occurred in 1820, 1875, and 1930. Appropriately, these troughs correspond to periods of pronounced recession, even depression, in the economic development of the United States.

From a historical perspective we can associate four technology clusters with this statistical pattern and speculate on the emergence of a fifth. The clusters may be identified by their most important economic branches, infrastructures, or functioning principles. Extending to the 1820s, we find textiles, turnpikes, and water mills; extending until about 1870 we find steam, canals, and iron; extending until about 1940 we find coal, railways, steel, and industrial electrification; extending to the present we find oil, roads, plastics, and consumer electrification (Grübler, 1994). Currently we appear to be in transition to a new era of industrial and economic development. We can speculate that it will be characterized by natural gas, aviation, "total quality control" of both the internal and external (or environmental) quality of industrial production, and the massive expansion of information handling.

These observations add up to an essentially Schumpeterian view of long-term development. Major economic expansion periods appear driven by the widespread diffusion of a host of interrelated innovations—a technology cluster—leading to new products, markets, industries, and infrastructures. These diffusion processes are sustained by, in fact are contingent on, mediating social and organizational diffusion processes. The growth or diffusion of a dominant cluster cannot be sustained indefinitely, however.

Market saturation, the dwindling improvement of possibilities for existing process technologies, managerial and organizational settings, and an increasing awareness of the negative (specifically, environmental) externalities involved in the further extension of the dominant growth regime pave the way to a season of saturations. During such periods, opportunities arise for the introduction of new technological, organizational, and social solutions, some of which may have been latent but were barred from market entry by the dominance of the previous growth paradigm. Even when such innovations are introduced successfully, their penetration rates in the initial phase of their diffusion life cycle are rather slow, and a matching new social and economic mediating context has still to emerge. In the phase-transition period, the old is saturating, and the new is still embryonic. Only after such a period of transition, crisis, and mismatch does a prolonged period of widespread diffusion of a new sociotechnical "bandwagon" and thus of growth become possible.

CONCLUSIONS

Empirical examination of diffusion processes, as illustrated in this essay, highlight the following observations:

(1) No innovation spreads instantaneously. Instead, a typical S-shaped temporal pattern seems to be the rule. This basic pattern appears invariant, although the regularity and timing of diffusion processes vary greatly.

(2) Diffusion is a spatial as well as temporal phenomenon. Originating from innovation centers, a particular idea, practice, or artifact spreads out to its hinterland by means of a hierarchy of subinnovation centers and into the periphery, defined spatially, functionally, or socially.

(3) The periphery, while starting adoption later, profits from learning and the experience gained in the core area and generally has faster adoption rates. As the development time is shorter, however, the absolute adoption intensity is lower than in innovation centers or in core areas (spatial or functional) proximate to them.

(4) Although diffusion is essentially a process of imitation and homogenization, it clusters and lumps. The densities of application remain discontinuous in time and heterogeneous in space among the population of potential adopters and across different social strata. In fact, overall development trajectories appear necessarily punctuated by crises that emerge in transitional periods. As such, diffusion and its discontinuities may be among the inherent features of the evolutionary process that governs social behavior.

Nevertheless, appropriate incentives and policies may nurture the development of more benign technologies and their diffusion, and many changes can be implemented over a time frame of two to three decades. However, sectors and areas will also remain in which changes will occur much more slowly, particularly those related to the long-lived structures of our built environment: for example, infrastructures for transport and energy as well as housing stock. Here rates of change and diffusion constants ranging from several decades to a century are typical and will be costly to accelerate. Therefore, the efficiency with which existing systems are used merits attention.

In essence we have two strategies in light of diffusion. One focuses on incremental changes, for example, environmental add-on or "end-of-pipe" technologies. Such policies can bring quick changes but tend to reinforce the dominant trajectory, blocking more systemic and radical changes. A second strategy opts for more radical departures from existing technologies and practices. However, these strategies, such as the development of fuel cells and hydrogen for energy, although more effective in the long run, require much more time to implement because of the multiplicity of forward and backward linkages between technologies, infrastructures, and forms of organization for their production and use.

The interdependence between individual artifacts and long-lived infrastructures creates our dilemma. Within two to three decades the United States could in principle change its entire fleet to zero-emission vehicles. In fact, 99 percent of vehicles now on the road will be scrapped in this interval. Yet, this interval is too short for the diffusion of the required associated energy supply, transport, and delivery infrastructures, which will inevitably distend the rate of diffusion of end-use devices. Thus, key technologies that we can already envision to raise the quality of the environment probably must await the second half of the twenty-first century to become widespread and influential.

Historically, technology clusters have been instrumental in raising productivity and also in alleviating many adverse environmental effects. The emergence of a new cluster could hold the promise of an environmentally more compatible technological trajectory. But it will take time. There are times of change and times for change, and unless our individual and collective behavior is modified, these times will remain to frustrate and excite us.

NOTES

1. For an extended version of this essay, see Grübler (1995).

2. On the spatial diffusion of Cistercians, see Donkin (1978). For a general overview of diffusion theory, see Hägerstrand (1967), Morill (1968, 1970), and Rogers (1983). For a more recent overview of diffusion theory, see Grübler and Nakićenović (1991). On the role of networks, see Kamann and Nijkamp (1991).

3. For an overview from sociology and anthropology, see Rogers (1983); for an overview from economics, see Mansfield (1961, 1968). For industrial innovations, see Nasbeth and Ray (1974) and Ray (1989).

4. For a more detailed discussion, see Freeman and Perez (1988) and Grübler (1992).

5. The equation to which the data are fitted has the form $Y = k/(1 + e^{(-b(t-t_m))})$, where $Y(t)$ represents the sigmoidal growth through time of a population or process, Y. This is often referred to as the logistic model. Three parameters control the shape of the sigmoidal growth trajectory: b controls the steepness (or diffusion rate) of the model; k denotes the asymptotic limit (or saturation level); and t_m denotes the middle or inflection point. The inflection point occurs at $k/2$, where the growth rate (dY/dt) is at a maximum. Note that k is sometimes also referred to as the "carrying capacity." A convenient notation for the diffusion rate (b) is Δt, where Δt is the time it takes for the process to grow from 10 to 90 percent of the saturation level, k. Approximately the same length of time is required for the process to grow from 1 to 50 percent. Through simple algebra, it can be shown that $\Delta t = \ln(81)/b$.

6. For an account of the dynamic interactions in US transport infrastructure development, see Nakićenović (1988). For a discussion of the impacts of transport infrastructure development on economic growth and discontinuities in economic development, see Berry et al. (1993), Grübler (1990), and Isard (1942). Berry (1990) also provides a good account of their impact on urbanization.

7. Starr and Rudman (1973) suggested a doubling time of twenty to thirty years for the technological component of economic growth, an estimate that our data sample corroborates.

8. For discussion of such distributions, see Montroll and Badger (1974).

REFERENCES

Arthur, W. B. 1988. Competing technologies: An overview. Pp. 590–607 in Technical Change and Economic Theory, G. Dosi, C. Freeman, R. Nelson, G. Silverberg, and L. Soete, eds. London: Pinter Publishers.

Basalla, G. 1988. The Evolution of Technology. Cambridge, England: Cambridge University Press.

Berry, B. J. L. 1990. Urbanization. Pp. 103–119 in The Earth as Transformed by Human Action: Global and Regional Changes in the Biosphere over the Past 300 Years, B. L. Turner, W. C. Clark, R. W. Kates, J. F. Richards, J. T. Mathews, and W. B. Meyer, eds. Cambridge, England: Cambridge University Press.

Berry, B. J. L., H. Kim, and H-M. Kim. 1993. Are long waves driven by techno-economic transformations? Evidence from the U.S. and the U.K. Technological Forecasting and Social Change 44:111–135.

Donkin, R. A. 1978. The Cistercians: Studies in the Geography of Medieval England and Wales. Toronto: Pontifical Institute for Mediaeval Studies.

Dosi, G., L. Orsenigo, and G. Silverberg. 1986. Innovation, Diversity and Diffusion: A Self-Organization Model. Sussex, England: Science Policy Research Unit (SPRU), University of Sussex.

Freeman, C., and C. Perez. 1988. Structural crises of adjustment, business cycles and investment behaviour. Pp. 38–66 in Technical Change and Economic Theory, G. Dosi, C. Freeman, R. Nelson, G. Silverberg, and L. Soete, eds. London: Pinter Publishers.

Grübler, A. 1990. The Rise and Fall of Infrastructures, Dynamics of Evolution and Technological Change in Transport. Heidelberg, Germany: Physica Verlag.

Grübler, A. 1991. Diffusion: Long-term patterns and discontinuities. Technological Forecasting and Social Change 39:159–180.

Grübler, A. 1992. Technology and Global Change: Land-Use, Past and Present. WP-92-2. Laxenburg, Austria: International Institute for Applied Systems Analysis.

Grübler, A. 1994. Industrialization as a historical phenomenon. Pp. 43–68 in Industrial Ecology and Global Change, R. Socolow et al., eds. Cambridge, England: Cambridge University Press.

Grübler, A. 1995. Time for a Change: Rates of Diffusion of Ideas, Technologies, and Social Behaviors. WP-95-82. Laxenburg, Austria: International Institute for Applied Systems Analysis.

Grübler, A., and N. Nakićenović. 1991. Long waves, technology diffusion, and substitution. Review 14(2):313–342.

Hägerstrand, T. 1967. Innovation Diffusion as a Spatial Process. Chicago: University of Chicago Press.

Herman, R., and E. W. Montroll. 1972. A manner of characterizing the development of countries. Proceedings of the National Academy of Sciences of the United States of America 69:3019–3023.

Hobsbawm, E. J., and G. Rudé. 1968. Captain Swing. New York: Pantheon Books.

Isard, W. 1942. A neglected cycle: The transport building cycle. Review of Economic Statistics 24(4):149–158.

Janauschek, P. L. 1877. Originum Cisterciensium. Tomus I. Vienna: A. Hoeler.

Kamann, D.-J. F., and P. Nijkamp. 1991. Technogenesis: Origins and diffusion in a turbulent environment. Pp. 93–124 in Diffusion of Technologies and Social Behavior, N. Nakićenović and A. Grübler, eds. Berlin: Springer-Verlag.

Mansfield, E. 1961. Technical change and the rate of imitation. Econometrica 29(4):741–766.

Mansfield, E. 1968. The Economics of Technological Change. New York: W. W. Norton.

Marchetti, C. 1980. Society as a learning system: Discovery, invention and innovation cycles revisited. Technological Forecasting and Social Change 18:267–282.

Marchetti, C., and N. Nakićenović. 1979. The Dynamics of Energy Systems and the Logistic Substitution Model. RR-79-13. Laxenburg, Austria: International Institute for Applied Systems Analysis.

Mensch, G. 1975. Das technologische Patt. Frankfurt: Umschau.

Montroll, E. W. 1978. Social dynamics and the quantifying of social forces. Proceedings of the National Academy of Sciences of the United States of America 75:4633–4637.

Montroll, E. W., and W. W. Badger. 1974. Introduction to Quantitative Aspects of Social Phenomena. New York: Gordon and Breach.

Morill, R. L. 1968. Waves of spatial diffusion. Journal of Regional Science 8:1–18.

Morill, R. L. 1970. The Spatial Organization of Society. Belmont, Calif.: Duxbury Press.

Nakićenović, N. 1986. The automotive road to technological change: Diffusion of the automobile as a process of technological substitution. Technological Forecasting and Social Change 29:309–340.

Nakićenović, N. 1987. Technological substitution and long waves in the USA. In Life Cycles and Long Waves, Lecture Notes in Economics and Mathematical Systems, T. Vasko, R. U. Ayres, and L. Fontvieille, eds. No. 340. Berlin: Springer-Verlag.

Nakićenović, N. 1988. Dynamics and replacement of U.S. transport infrastructures. Pp. 175–221 in Cities and Their Vital Systems, Infrastructure Past, Present, and Future, J. H. Ausubel, and R. Herman, eds. Washington, D.C.: National Academy Press.

Nasbeth, L., and G. F. Ray, eds. 1974. The Diffusion of New Industrial Processes: An International Study. Cambridge, England: Cambridge University Press.

Pemberton, H. E. 1936. The curve of culture diffusion rate. American Sociological Review 1(4):547–556.

Ray, G. 1989. Full circle: The diffusion of technology. Research Policy 18:1–18.

Reynolds, R. V., and A. H. Pierson. 1942. Fuel Wood Used in the United States, 1630–1930. Circular 641. Washington, D.C.: US Department of Agriculture.

Rogers, E. 1983. Diffusion of Innovations. Third Edition. New York: Free Press.

Schumpeter, J. A. 1939. Business Cycles, A Theoretical, Historical, and Statistical Analysis of the Capitalist Press. Vols. I and II. New York: McGraw-Hill.

Silverberg, G. 1991. Adoption and diffusion of technology as a collective evolutionary process. Pp. 209–229 in Diffusion of Technologies and Social Behavior, N. Nakićenović and A. Grübler, eds. Berlin: Springer-Verlag.

Silverberg, G., G. Dosi, and L. Orsenigo. 1988. Innovation, diversity and diffusion: A self-organisation model. The Economic Journal 98(December):1032–1054.

Starr, C., and R. Rudman. 1973. Parameters of technological growth. Science 182(26):360.

van Duijn, J. J. 1983. The Long Wave in Economic Life. London: Allen and Unwin.

Vasko, T., ed. 1987. The Long-Wave Debate. Berlin: Springer-Verlag.

Population, Technology, and the Human Environment: A Thread Through Time

ROBERT W. KATES

Plato observed it, the old testament taught it, and Thomas Robert Malthus feared it. It has been called the principle of plenitude, which "presupposes a richness, an expansiveness of life, a tendency to fill up, so to speak, the empty niches of nature; implicit is the recognition of the great variety of life and perhaps its tendency to multiply" (Glacken, 1967, p. 57). For all living things, the biblical injunction is clear: "Be fruitful and multiply" (Genesis 1:22). But for one species of life, humans, the injunction is clearer yet: "Be fruitful and multiply, and fill the earth and subdue it; and have dominion over the fish of the sea and the birds of the air and over every living thing that moves upon the earth" (Genesis 1:28).

Malthus, a Christian cleric, worried over the injunction and conducted a thought experiment to demonstrate how disastrous its pursuit would be:

> . . . if the necessaries of life could be obtained and distributed without limit, and the number of people could be doubled every twenty-five years, the population which might have been produced from a single pair in the Christian era, would have been sufficient, not only to fill the earth quite full of people, so that four should stand in every yard, but to fill all the planets of our solar system . . . and the planets revolving around the stars which are visible to the naked eye (Glacken, 1967, p. 641).

Thus, Malthus concluded, a benevolent Creator would limit in quantity "the necessaries of life" and temper the principle of plenitude by the principle of population.

Malthus's principle of population begins with the living realities of hunger and sex, and the latter can be satisfied with greater ease than the former. Sex in

Malthus's time was still linked to frequent reproduction, leading to a faster growth in the numbers of persons than in the means of subsistence. Unchecked, "the human species would increase as the numbers 1, 2, 4, 8, 16, 32, 64, 128, 256; and the subsistence as 1, 2, 3, 4, 5, 6, 7, 8, 9." The imbalance cannot continue, and indeed growth is reduced by "positive checks" in the form of misery (famine, war, and disease) and vice (prostitution, homosexuality, adultery, birth control, and abortion) and—in revisions of his original *Essay on the Principle of Population*—by "preventive checks," primarily, delayed marriage.

Malthus was born in 1766, late in a century in which England and Wales almost doubled in population from 4–5 million to 9–10 million. Yet public controversy about human numbers raged in British intellectual circles until the first census of 1801, with many believing that Britain was losing population while it actually gained.

Educated by his father and tutors of independent mind, Malthus entered Cambridge in 1784 and graduated with honors in mathematics (James, 1979; Petersen, 1979). Like many scientists and intellectuals of his generation, he became both a university fellow and an Anglican priest. He published his *Essay* anonymously in 1798 at the age of thirty-two, while serving as curate at a small country chapel. In Surrey, in the village of Oakwood, Malthus presided over numerous baptisms and may have directly observed the rapid growth of the English population. Death was also known, and Adam Smith, the most powerful intellectual influence on Malthus, had written in *The Wealth of Nations* (1776) that

> . . . in some places one-half of the children born die before they are four years of age; in many places before they are seven; and in almost all places before they are nine or ten. . . . Every species of animals naturally multiplies in proportion to the means of subsistence, and no species can multiply beyond it. But in a civilized society it is only among the inferior ranks of people that the scantiness of subsistence can set limits to the further multiplication of the human species and it can do so in no other way than by destroying a great part of the children which their fruitful marriages produce (Petersen, 1979, p. 40).

Still, the sources of inspiration for Malthus are not obvious. Of course, other limits to population had been observed. Some 1,600 years previous, Tertullian, a Carthaginian resident in Rome, wrote:

> Surely, it is obvious enough, if one looks at the whole world, that it is becoming better cultivated and more fully peopled than anciently. . . . No longer are savage islands dreaded, nor their rocky shores feared; everywhere are houses, and inhabitants, and settled government, and civilized life. What most frequently meets our view is our teeming population; our numbers are burdensome to the world, which can hardly supply us from its natural elements; our wants grow more and more keen, and our complaints more bitter in all mouths, whilst nature fails in affording us her usual sustenance. In very deed, pestilence and

famine, and wars, and earthquakes have to be regarded a remedy for nations, as a means of pruning the luxuriance of the human race (Brown, 1954, p. 30).[1]

Firmly setting one pole of a profound disagreement that persists to this day, Malthus published partly in response to the utopian visions of human perfection offered by his contemporaries, such as William Godwin and the Marquis de Condorcet (see Meyer-Abich, this volume). Against their confidence in human institutions and ingenuity, Malthus invoked the hard arithmetic and biological and environmental determinism of the principle of population.

All of us who ponder the questions of the human environment are the intellectual descendants of Thomas Robert Malthus. Whether "neo-Malthusian," "anti-Malthusian," or simply agnostic, we explore the equation of population with resources and technology, which distills the problem of the human environment. Over time the focus of Malthusian concerns has shifted. In 1798 the key ratio in the Malthusian equation was food and farmland per person. By the 1850s, the resource term expanded to include energy and other materials, urgently argued in the classic volume of British economist William Jevons on the coal question (Jevons, 1865). By the middle of the twentieth century, the United States would discount fears about resource scarcity and promote a new Malthusian numerator that included amenity resources and the pollution-absorbing capacity of the environment (President's Materials Policy Commission, 1952). The UN Stockholm Conference on the Environment in 1972 enlarged such concerns to a global scale and drew attention to the basic life-support systems and the chemical cycles of the biosphere. More recently, losses in the diversity of life and genetic information have joined the earlier concerns.

Characteristically, none of the earlier Malthusian concerns really disappear but are renewed in some larger, more international context. And for each of the different notions of critical resources, technology will make possible new reserves and new substitutions and in turn cause new problems. Thus, a continuous process of Malthusian refutation and renewal has marked the two centuries since publication of the *Essay*. In my own professional life, I have participated in two and a half cycles of research and argument. Currently, I am trying to understand the roles of neo-Malthusian scientific "Jeremiahs" and society's response to them by examining the post-World War II history of jeremiads (Kates, 1995), beginning with Vogt's (1948) and Brown's (1954) concerns with population growth, moving on to subsequent fears about food, materials, energy, and toxic pollutants, and concluding with the formal synthesis of concerns in *The Limits to Growth* (Meadows et al., 1972).

Over time, the population denominator has increased from a local to a national, regional, and then global scale. The requirements of each person also change over time, from the meager demand typical of Malthus's day to the copious consumption of the wealthy fifth of the present world population (see Wernick et al., this volume). Contrasts with the modest per capita usage of most residents of the less-industrialized countries show how levels of affluence and types of technology modify the Malthusian equation (Ehrlich and Ehrlich, 1990).

Yet beginning with the genus *homo,* the numbers of people form a continuous thread through time with which to examine the warp and woof that pattern our environment. Thus, in this essay, I employ a sequence of four temporal frames—ages, millennia, centuries, and decades—to examine the dynamics of population, resources, and technology. Each frame highlights significant questions about the sources of technological change; the growth, decline, and stabilization of human populations; and the extraordinary challenge posed by the dynamics of the current period.

AGES: TECHNOLOGICAL REVOLUTIONS AND POPULATION SURGES

A few years ago, I had the opportunity to review *Beyond the Limits* (Meadows et al., 1992), the sequel to *The Limits to Growth* (Meadows et al., 1972). In the words of the authors, "human uses of many essential resources and the generation of many kinds of pollutants have already surpassed rates that are physically sustainable" (Meadows et al., 1992, p. xv). The first figures in the book contain the familiar curves of exponential growth over several centuries of world population and, during this century and more fitfully, industrial production (see Figures 1 and 2). These the authors generalize, stating that "Exponential growth is the driving force causing the human economy to approach the physical limits of the earth" (Meadows et al., 1992, p. 14).

While I suspect that many who casually encountered that statement might agree with it, I experienced a deep uneasiness with this frequently-used mental graphic of the future. Having assimilated, as a graduate student, a different image of population pathways, past and future, as S-shaped curves growing to limits, I was left forever skeptical of the exponential vision. In a 1960 article, ecologist Edward Deevey had pointed out two defects in the commonly accepted picture of the growth of the population shown in Figure 1. First, the basis of the estimates, back to about A.D. 1650, is rarely stated. Second, the scales of the graph are chosen so as to make the first defect unimportant. In Deevey's words, "The missile has left the pad and is heading out of sight" (Deevey, 1960, p. 197).

To remedy this situation, Deevey collected the then-available estimates of population over hominid existence and plotted these on logarithmic scales to emphasize ratios rather than absolute numbers. These curves and more recent data are shown in Figure 3. Deevey explained that:

> The stepwise evolution of population size, entirely concealed with arithmetic scales, is the most noticeable feature of this diagram. For most of the million-year period, the number of hominids, including man, was about what would be expected of any large Pleistocene mammal—scarcer than horses, say, but commoner than elephants. Intellectual superiority was simply a successful adapta-

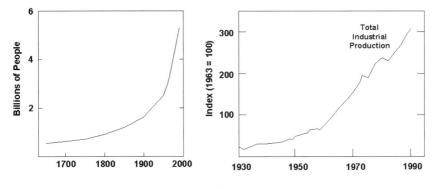

FIGURE 1 World population.   **FIGURE 2**   World industrial production.

tion, like longer legs; essential to stay in the running, of course, but making man at best the first among equals. Then the food-gatherers and hunters became plowmen and herdsmen, and the population was boosted by about sixteen times, between 10,000 and 6,000 years ago. The scientific-industrial revolution, beginning some 300 years ago, has spread its effects much faster, but it has not taken the number as far above the earlier baseline. The long-term population equilibrium implied by such baselines suggests something else. Some kind of restraint kept the number fairly stable (Deevey, 1960, pp. 197–198).

According to Deevey, human population has surged greatly three times. The first was associated with the toolmaking or cultural revolution, lasted about a million years, and saw human numbers rise to five million. The second saw the population swell a hundredfold to about five hundred million people over the next eight thousand years, following the domestication of plants and animals and the invention of agriculture and animal herding. In Malthus's lifetime, early in the industrial revolution, it doubled again to the first billion. With a current world population of 5.7 billion, we are in the midst of the final doubling of this, the third great surge of the population. World population is projected to increase to more than eleven billion before leveling off again—some three to four hundred years after the scientific-industrial revolution began.

These toolmaking, agricultural, and scientific-industrial revolutions each transformed the meaning of resources and increased the carrying capacity of Earth. Each made possible a period of exponential growth followed by a period of approximate stability, as the record of human existence reveals in the frame of ages.

But if this is the record, the causes of such technological change are not as clear. Consider, for example, the origins of agriculture. Intentional farming for food or subsistence dates back nine thousand years. Agriculture evidently began independently in the Near East between eight thousand and nine thousand years

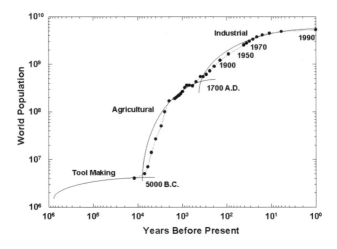

FIGURE 3 World population with three growth pulses. NOTE: Updated and redrawn from Deevey (1960). SOURCES: Deevey (1960), McEvedy and Jones (1985), and United Nations (1993).

ago for wheat and barley, eight thousand to seven thousand years ago in China for millet and rice, and eight thousand to sixty-five hundred years ago in the Western Hemisphere for squash and maize (Cowan and Watson, 1992; Mathews et al., 1990; Reed, 1977). Agriculture and its pastoral cousin gradually replaced a system of food gathering and hunting that had apparent advantages of less work and better diets (Cohen, 1977, 1990; Sahlins, 1972). Why? The short answer is, "We do not know"; the longer one begins, "We have theories."

In general, the many explanations emphasize either push or pull factors. The pushes to agriculture are primarily said to be population increase and environmental change. Human communities six thousand to nine thousand years ago turned to agriculture because their numbers increased beyond the carrying capacity of their accessible resource base, or the resource base was reduced by environmental (climatic, biological, or human-induced) changes, or both. The pulls highlight the attractiveness of agricultural technology (or the agrarian life-style) in increased yields per hectare and in the ability to store resources, thereby reducing annual and seasonal variation in the food supply. Thus, human communities encountered wild precursors of domesticated plants and animals; gradually learned about their availability, reproduction, and life cycle; and then experimented, intentionally or incidentally, with their selection, growth, harvesting, and use.

Within push and pull there are many variants as well as hybrid explanations that emphasize one factor or another in a dynamic sequence. A coevolutionary explanation even argues against the independence of human agency implied by both push and pull theorists (Rindos, 1984). Instead, it offers the perspective of the domesticated plants and animals and their seeming reproductive success by encouraging humans to domesticate them, a quite different view of humans in nature (see Meyer-Abich, this volume).

Influenced by the Danish economist Ester Boserup (1965, 1981), demographer Ronald Lee attempts to transcend the particular explanations for each of the great Deevey revolutions by integrating the theoretical insights of the

> ... two grand themes in macro-demographic theory: the Malthusian one, that population equilibrates with resources at some level mediated by technology and a conventional standard of living, and the Boserupian one, that technological change is itself spurred by increases in population. The striking association between the levels and changes in technology and population over the past million years leaves no doubt in my mind that at least one of these views is correct. But it is also possible that both are, since the two theories are not contradictory, but rather complementary. They share the assumption of diminishing returns to labour for a fixed technological level. To this common ground Malthus adds the assumption that population growth rates are endogenous, while Boserup adds the assumption that technological change is endogenous (Lee, 1986, p. 96).

Lee develops the broad qualitative features of a dynamic system governed by the mechanisms of both Malthus and Boserup and applies it to the Deevey diagram, asking how the transition between technological revolutions might be made. Lee defines a Malthus space in which, for a given level of technology, population grows; and a Boserup space in which, for a given level of population, technology grows:

> ... for any state of technology, there are some ranges of population size within which technological progress occurs, and others where it does not. For any given state of technology, Malthusian forces will steer population size towards some equilibrium level. Common sense suggests that the behaviour of the system will depend critically on whether this Malthusian equilibrium population size falls within the range leading to further technological advance, or within the range leading to technological regression, either because the population is too small or too large (Lee, 1986, p. 118).

Lee sees the great technological revolutions as three distinct or weakly connected domains, each constrained by Malthusian equilibria.

To explore the theory, Lee considers the ability of cultures to leap across to other distinct technological regimes or pass through the bottleneck of weakly connected ones. He addresses the puzzling failures of China, more technologically advanced than Europe, to move early into the industrial revolution, and of Africa to move beyond hoe agriculture. These, Lee speculates, might be explained by Africa having too few people to force needed levels of technological intensification. And China, with too many people to accrue the surplus needed to invest in the crucial technology, perhaps found itself limited to mid-level technologies that the Europeans, with greater investment, repeatedly improved upon.

The complex and seemingly endless discussions on the origins of agriculture in archaeology, anthropology, biology, demography, economics, and geography, and the effort by Lee to develop an integrative theory, suggest two conclusions. Intense study has not yielded ready, simple, or consensual explanations as to the causes of the great technological revolutions. The most credible explanations depend on historical detail, multiple causes, and dynamic forces. They also yield a question about the driving forces of technological trajectories, such as decarbonization and dematerialization, documented elsewhere in this volume (see Nakićenović and Wernick et al., this volume). Is there some push of necessity that drives these forces, some teleological pull of technological superiority or economic efficiency, or some coevolutionary process of the natural selection of technologies within the human environment?

MILLENNIA: WHAT GOES UP MAY COME DOWN

Although the graphic message of the ages for the entire Earth is three great logarithmic arcs, the message of smaller frames differs. Reconstructing the population of regions over thousands of years, we find what my colleagues and I have called "millennial long waves." It is not surprising, after all, that societies might have some long harmonics, considering the range of time scales reported for the diffusion of various ideas and technologies (see Grübler, this volume).

These population reconstructions grew out of an effort to examine how the time scales of human societies match other processes in nature, with many cycles of human activities contained in lifetimes or generations and those of the environment extending also to centuries and ages. We sought to compare the longest continuous place-based sequences of human activity that we could construct and to relate these, in turn, to environmental change. We were able to do this for four regions: the Egyptian Nile Valley, the Tigris-Euphrates lowlands, the Basin of Mexico, and the central Mayan lowlands of Mexico and Guatemala (Whitmore et al., 1990).

These regions range in size from the compact Central Basin of Mexico of about 7,000 km^2 to the extensive Tigris-Euphrates lowlands of about 55,000 km^2. Their durations span the six-thousand-year reconstruction of the Nile Valley and the three thousand years of the central Mayan lowlands. The area of each region was selected on the basis of the congruence between a particular culture and a distinctive physical environment in the earliest period of the reconstruction and was then kept constant through the entire reconstruction. The duration of each reconstruction was based on the ability to meld archaeological and historical data to create a long-term sequence of estimated population. Methods involved the conversion of both archaeological material (e.g., ceramic or habitation remains)

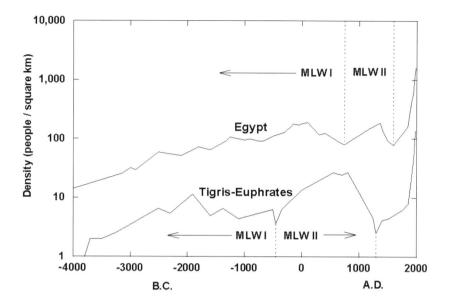

FIGURE 4   Tigris-Euphrates and Egyptian populations.   NOTE: MLW = millennial long wave.

and documentary (tax or census) records into site-specific population estimates. Because for the most part we used estimates drawn from the work of other researchers, we selected between competing estimates based on our judgment of their demographic probability, quality of source data, and the validity of estimation techniques employed. Where needed, we inferred missing values for key time intervals.

The reconstructed population series are shown in Figures 4 and 5. They evidence both growth and decline; in none is population growth simply upward and onward from the cave. To highlight and compare major episodes of growth and decline and to distinguish these from fluctuations that were minor or artifacts of the estimation methods, we adopted a convention of considering only variations in growth in which the population at least doubles from its preexisting base, or is in decline in which it is minimally halved from its intervening apex. (This is akin to the risk assessment convention that considers as risk factors only those that at least double the observed risk.) With this criterion, each record is divided into intervals that we have designated as millennial long waves (shown as MLW I or MLW II in Figures 4 and 5).

In all except the Mayan case (the shortest record), the reconstruction shows two waves in which the population at least doubled over the previous base and then at least halved from that high point, as well as the rising part of a third wave.

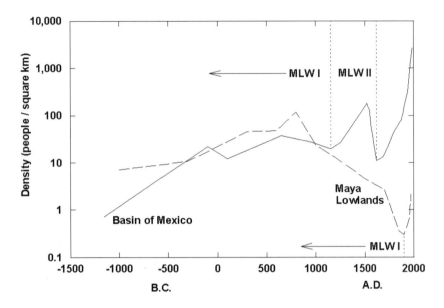

FIGURE 5   Mexican and Mayan populations. NOTE: MLW = millennial long wave.

While the waves are all very long, they decrease in duration. The first waves average about 3,600 years in length, the second about 1,500 years, and the growth phase of the third waves, still in progress, averages 380 years to date. Growth phases last longer, occupying about 70 percent of the reconstructed time period. Rates of growth increase over time, averaging for the first waves 0.14 percent per annum, for the second 0.30 percent per annum, and for the modern period 1.43 percent per annum. The decline phases, while shorter and surely catastrophic, are not exactly precipitous. The second waves, for example, average more than five hundred years in duration even though they include one of the most precipitous population drops in human history—the sixteenth-century die-off of the native peoples of the Americas—whose immediate cause was epidemics of infectious disease.

What drives such long waves of increasing frequency and great amplitude? Again, we do not know, but we have theories. For one case (Bowden et al., 1981), the Tigris-Euphrates lowlands, we compiled a parallel reconstruction of major social, technological, and environmental events. We found no simple correlation between population growth and decline in the Tigris-Euphrates flood plain and periods of state formation, war, and empire collapse, or technological or climatic change. Rather, the interaction of the social, technological, and environmental events may cause the long-term population growth and decline. A simulation model has plausibly reproduced some of those interactions (Johnson and Gould, 1984).

The long waves of growth and decline disappear at the global scale of ages, as seen in the graph of logarithmic growth and stabilization (see Figure 3). Presumably the fate of particular places is averaged out. Some grow, others decline, but the overall tendency is growth. Has the scientific-industrial revolution, with a global economy and a global famine response system, exempted us from the Malthusian collapses of the past? Or can the collapse of particular regions, including regions that are world leaders, still occur in the modern world?

The millennial perspective offers no encouragement for an exemptionist doctrine. Indeed, by way of a thought experiment, I have tried to develop reasonable, albeit imaginative, decline scenarios for the current third wave still in its growth phase. The growth phase is projected to end in the 2060–2080 period, a time when current long-term demographic projections find that the relevant national populations will have accumulated 95 percent of their hypothetical equilibrium population. The duration of the decline of the third wave is estimated to be 0.65 of the growth phase, based on historic ratios. An average decline period of about three hundred years would then follow the 2060–2080 peak. We could consider, therefore, scenarios such as these:

Egypt: The Nile Valley population peaks in 2080 at about 110 million people then begins a sharp decline. Three factors contribute to the decline: the development of a mechanized agriculture outside the valley that competes for Nile water; the suburbanization of Cairo; and most importantly, the recurrent bouts of MAIDS fever, the molluscan autoimmune disease.

Tigris-Euphrates: The city of Baghdad and the dams and weirs of the Tigris-Euphrates are targeted in the second war with the Elamite Democratic Republic and are never rebuilt.

Basin of Mexico: Repeated attempts by six successive Mexican governments to decentralize government, industry, and services outside the Basin of Mexico fail, and Mexico City becomes the largest city in the world. However, a succession of disasters—beginning with the Great Vulchemical Smog of 2112 and ending with the Earthquake of 2119, which leaves 35 percent of the buildings uninhabitable—finally leads to the relocation of the capital to the site of the ancient city of Monte Alban, 250 miles southeast.

Mayan Lowlands: Clearing the central Mayan lowlands has newly revealed two major ancient urban sites at Uaxtum and Real Azul as well as reduced the habitat of rare birds. Through an initiative of the Organization of Central American States, the first trinational archaeological and biological park in the world is created for tourism, research, and wildlife conservation. In all, 22,715 km^2 are purchased in Belize, Guatemala, and Mexico and set aside for this purpose. Beyond the required staff no permanent inhabitants reside in the park after relocation.

CENTURIES: FROM HIGH BIRTHS, HIGH DEATHS
TO LOW BIRTHS, LOW DEATHS

In the 1760s, the decade of Malthus's birth, England and Wales grew by 7 percent; in the 1790s, the decade of the first edition of his *Essay,* they grew by 11 percent. Three editions later, as Malthus neared death in 1834, decadal population growth had peaked at 18 percent—a veritable explosion (Clark, 1968). The driving force behind the growth was primarily the decline in death rates. Thus, to have a decline in growth, birthrates also had to decline. Even as Malthus wrote the first *Essay,* the birthrate had declined in France by about 10 percent. Yet it would take almost a century more to decline in England and Wales and more than a century for all of Europe (Knodel and van de Walle, 1976). An emigration of fifty million people was also important in balancing European births and deaths. This transition from high births and high deaths to low births and low deaths took about 150 years to complete in England and has become the classic episode for population study. Demographers call it, not surprisingly, "the demographic transition" (Davis, 1990).

Yet here, too, much is unknown or in dispute. The transition clearly coincided with a profound shift from a rural agrarian society to an industrialized economy. Conditions changed drastically, particularly for infants and children, and wealth and education increased. But the reasons why the number of deaths declined when and as they did are not truly understood (Walter and Schofield, 1989). In our age of medical miracles, it is easy to presume that improved health care and prevention, including immunization, resulted in the decline. But the death rate began to decline long before the knowledge and practices of modern medicine evolved (McKeown, 1976).

The postponement of death and the concomitant transition from high death rates to low death rates seem related to three different sources of disease (Kunitz, 1986). Before a society advances far in the demographic transition, most people die from infectious disease, although some diseases are intimately related to hunger. Early in the transition, deadly epidemics of infectious disease are moderated or disappear. These are diseases that you "catch" from food or water, or from another person, a rat, or a fly. The infectious agent may not normally live in our midst but, when present, can infect both the well-fed and the hungry. Such diseases include bubonic plague, malaria, measles, smallpox, typhus, and yellow fever. They recede in the face of quarantine, spraying, vaccination, sanitary practices, and, to some degree, through a growing immunity within the population. Thus some diseases, measles and chicken pox, become children's diseases; the rest of us have already had them. In England this part of the decline began as early as the 1670s, when plague seemed to disappear.

More stubborn are the infections from the endemic diseases, for which the sources of infection are ever present. Pneumonia, tuberculosis, and diarrhea are propagated by malnutrition, crowding, and poor sanitation. Progress against these diseases requires improvements in diets and living conditions.

Finally, deadly infections are replaced by the noninfectious diseases, the so-called diseases of civilization: heart disease and cancer. But by this time the death transition is over, and these are diseases of aging. Life expectancy is at least seventy years, and 1 percent or less of the population will die each year.

The decline in births lagged behind the decline in deaths, but in Britain both changes followed industrialization, as the society moved from an agrarian to an urban base. Thus, many scholars associate the decline in births and deaths with modernization or development (Davis, 1990). Scholars differ on which elements of development would encourage the decline in births—the changing economics and usefulness of family labor, the improved security of family size with reduced infant and child deaths, or greater knowledge and interest in birth control resulting from education, particularly of women. Whatever the reasons, fertility began its decline in most European countries between approximately 1880 and 1930, first in Belgium and ending in Ireland (Knodel and van de Walle, 1976).

In many countries, including Belgium, England, Germany, and Switzerland, the development process was well under way when birthrates began to decline, lending support to the theory. But in others, including Bulgaria, Hungary, Italy, and Spain, birthrates turned downward while the societies remained predominantly agrarian and illiterate. Indeed, France began its decline before its industrial revolution. And in all European countries, births turned downward while infant mortality remained high, as high as the highest rates anywhere today or higher; so high rates of child survival were not a prerequisite of the decline.

While the empirical facts of the demographic transition are clear, the causes—even in the best-studied historical cases—are not. In the demographic transition now under way among the three-quarters of the world population found in developing countries, opportunities exist for both better understanding and further complication by virtue of the added conscious effort to influence the transition, a factor absent in the case of Europe.

For nations now in demographic transition, the process is also more rapid. The United Nations' first long-range projections of world population forecasted a population of 3.8 billion by 1975 based on medium assumptions (United Nations, Department of Economic and Social Affairs, 1958). Underlying the UN projection was an expectation that the birthrate in 1975 would be 37 for each 1,000 persons and the death rate 17 per 1,000. The actual population in 1975 was 4.1 billion, close enough, but births stood at 30 per 1,000 and deaths at 12. Both birth- and death rates had dropped faster than experts expected and history foreshadowed. It took a hundred years for deaths to drop in Europe, whereas the drop took thirty years in the Third World. Today the global transition to the level required for stability is more than halfway between the average of five children born to each woman during the mid-century height of population increase and the

2.1 births that would eventually achieve zero population growth. Current births average about 3.1.

The global death transition is more advanced. Life expectancy has transited more than two-thirds of the way between a life expectancy at birth of forty years to one of seventy-five, and it is currently sixty-six years. What do we know about the causes of this transition? Experts widely agree on the effectiveness of control of the epidemic and endemic diseases and overall improvement in nutrition, sanitation, and public health in the developing countries. Life expectancy in high-death countries rises most rapidly from improvements in child survival; these have been quick, impressive, and, once begun, seem to continue even during stagnation in economic and social development. Much of the improvement has been intentional, a conscious application of nutritional and public health measures in developing countries using modern health care research and disease control technology. Unlike much current medical technology, applications such as immunization, diarrheal control, malarial control, smallpox eradication, and child nutrition have been both effective and relatively low cost.

It is characteristic to be in the midst of change and not recognize it. As mentioned, while Malthus wrote his *Essay,* his contemporaries debated whether the population of England was growing or declining. So, on the eve of the 1974 World Population Conference in Bucharest, a leading demographer, Ansley Coale, found little evidence of a fertility decline in the developing world except in several small countries with populations of Chinese origin (Coale, 1973). We now know that birthrates decreased in the decade from 1965 to 1975 by about 13 percent, with declines occurring in 127 countries (Population Reference Bureau, 1976). By the time of the conference, the birth decline phase of the Great Transition was already under way.

The conference brought together representatives of 136 countries. The United Nations had declared 1974 to be World Population Year, and the Bucharest meeting capped it. The meeting, despite the polite consensual rhetoric of its final statements, showed a profound split between the First World of industrialized countries, on one side, and the Third World of developing countries allied with the Second World of socialist countries, on the other (Finkle and Crane, 1975). Most conferees agreed on the need for a decline in population growth but split in their assessment of the requirements for the transition. A phrase of an Indian delegate, "Development is the best contraceptive," was the rallying cry for the Third World countries. Lack of development encouraged large family sizes, and social and economic development would bring, as it had in Europe, a decline in fertility and population growth, even without organized government population programs. Arrayed against these arguments were most Western European countries (except France and Italy), Canada, the United States, some Latin American countries, Australia, Japan, and Iran. While acknowledging the need for development, they advocated independent and organized efforts to reduce fertility and argued that such successful efforts would in turn lead to development itself.

Implicitly, the nations argued over two of the three major explanations for the rate and timing of the demographic transition: was it development or the access to modern knowledge and techniques of contraception that reduced family size?

A third major causal factor remained unspoken—culture and ethnicity. Such differences, whether real or not, were not discussible within the confines of the United Nations. Coale and others had noted, however, that the transition was most noticeable in countries populated by Chinese or those of Chinese origin, suggestive of the anomalous and pioneering role of France in the transition in Europe. Feminism was also largely unrecognized at the Bucharest Conference, as was the impact of changing education, employment, and roles for women.

At the September 1995 UN Cairo sequel to Bucharest, experts, including many women, demonstrated how such changing roles contribute to a fertility decline—although, as with the European transition, much still puzzles us. For example, it is not much clearer today which aspects of development most encourage lower birthrates in Africa, Asia, and Latin America than it was in the European decline. Analysts now choose from at least four arguments:

Less need for child labor, more need for educated children. As a society shifts from rural agrarian to urban industrialized, the potential contribution of children to family welfare and costs changes. The need for child labor lessens as does the role for children in providing old age security. Parents also make bigger investments in each child's health and education and expect greater returns to those investments in their future earnings. More care, energy, and money is spent on fewer children.

Less need for more births because more children survive. As the death transition proceeds, families realize that they can have the desired family size with fewer births since the chances of children surviving have increased.

Less time for childbearing and rearing, more time and need for education and work. As opportunities improve for women to have access to education and to work outside the household, marriage is delayed, and fewer births result from each marriage. Education and work compete for time with childbearing and provide alternative sources of reward and esteem.

More access to birth-control technology to achieve fewer births. Widely available, adequate, low-cost technology helps control the timing of conception. Access to such technology fulfills often long-standing desires for smaller families.

Of course, changing needs for labor, greater child survival, improved opportunities for women, and access to birth control all seem to proceed together in the course of development.

The protagonists in Bucharest and Cairo, however, cared less for the details of development than the distinction between development and organized family-planning efforts. Thus, much research has focused on seeking to estimate the relative contributions of economic and social development and organized family planning programs to the decline in births.

In comparing countries or regions, measuring development and characterizing family-planning programs are difficult. Even harder is disentangling the effects of development and organized family-planning programs since obviously they are strongly related. Development encourages people to use family-planning services. Indeed, organized family-planning programs are part of development, a natural occurrence in the provision by modern societies of health and welfare programs. Also, development creates the skilled people, transportation, access points for services, funding, and overall efficiency needed for effective programs. In turn, the results of effective family-planning programs might, over time, contribute to further development (Hernandez, 1981).

Attempting to control for these interactions, several cross-cultural studies, covering ninety-four or more countries (Lapham and Maudlin, 1984, 1987; Maudlin and Berelson, 1978; Sherris, 1985), have found that increases in development are strongly associated with a decline in the birthrate and in fact account for about two-thirds of the decline. And over and above development—or even the way development makes programs more effective—organized family-planning programs make an additional difference of 15–20 percent. But even this amount is disputed, with other analysts claiming that at most 5 percent of the fertility decline results from such efforts (Hernandez, 1981; Puitchett, 1994).

Only a few studies include other factors of culture and ethnicity. Yet if one looks further at the first ninety-four countries studied, taking the top twenty that recorded 20 percent or greater declines in births (compared with the overall world average of 13 percent), almost half the countries are in East or Southeast Asia, and a quarter are in the Caribbean. Of the top twenty, more than half are small island or city states. By numbers of population, Chinese speakers in China, Taiwan, Hong Kong, Singapore, and Malaysia predominate. Thus, one might add that when socioeconomic development and substantial family-planning programs are carried out in East and Southeast Asia, on small and crowded island or city states, or among those of Chinese extraction, more rapid declines take place.

Understanding the causes of fertility decline is not simply a scholarly undertaking but a pressing concern since the transition may have stagnated in the last decade. In fifteen countries, thirteen of them in Africa, birthrates apparently rose between the 1960s and the 1980s. In another twenty-three countries, the birthrate fell by less than 2 percent. In the 1970s total fertility dropped by 14 percent worldwide, in the 1980s by less than half that rate (Sadik, 1991). Both China and India had recent censuses and found higher populations than projected: seventeen million in China and four million in India. However, the recent Nigerian census found many fewer people than anticipated.

Cutting the average number of children that women bear from six to four has proven relatively easy in many developing countries. Further reduction, however, has been hard. The reasons may involve reduced political support in some countries (especially in the Near East, with the rise of religious fundamentalism); reduced spending because of debt-related cutbacks in health, education, and family planning; and the general slowing of development through the widespread economic stagnation of the 1980s and 1990s. Even more important, given the limited confidence in forms of social security outside the family unit, four may be the number of children actually desired in many parts of the developing world, and much of the previously unmet need may have now been met (Sadik, 1991). Finally, some scientists believe in African exceptionalism—that in Africa cultural, religious, and economic reasons encourage high fertility rates as much as East Asia seems to favor reductions in fertility (Caldwell and Caldwell, 1987). Countering these trends is the renewed momentum to limit births in China (Peng, 1993), promising changes in fertility decision-making in South India (Caldwell et al., 1988), and the first significant drops in fertility in several Southern and Eastern African countries (Caldwell, 1994).

DECADES: THE CHALLENGE OF THE GREAT CLIMACTERIC

We may well be in the final phase of the demographic transition of the scientific-industrial revolution, but from the perspective of the decades ahead, this is surely the Great Climacteric. At least that is how Ian Burton and I viewed it a decade ago:

> A climacteric . . . is a "critical period of human life" and a "period supposed to be specially liable to change in health or fortune" (*Oxford English Dictionary*). The term is normally applied to the individual; but as applied to population, resources, and environment throughout the world, it aptly captures the idea of a period that is critical and where serious change for the worse may occur. It is a time of unusual danger (Burton and Kates, 1986, p. 339).

In an extraordinarily short interval—a matter of decades—human society will need to feed, house, nurture, educate, and employ as many more people as already live on Earth. For this task, Deevey's interpretations of the past provide little comfort. A hundredfold increase in population marked past technological revolutions. The current multiplication is projected to be only two or perhaps three times, but we travel the trajectory within the span of a human lifetime.

Notwithstanding a wide range of estimates of how many people Earth can support (Cohen, 1995), for many of today's Jeremiahs a world of more than five billion people is already overpopulated. Ecologists Anne and Paul Ehrlich assert:

> The key to understanding overpopulation is not population density but the numbers of people in an area relative to its resources and the capacity of the environment to sustain human activities: that is, to the area's carrying capacity. When is an area overpopulated? When its population can't be maintained without rapidly depleting nonrenewable resources (or converting renewable resources into nonrenewable ones) and without degrading the capacity of the environ-

ment to support the population. . . . *By this standard, the entire planet and virtually every nation is already vastly overpopulated* (Ehrlich and Ehrlich, 1990, p. 38).

Many of us believe that if population growth can be held to some reasonable number, then sufficient food can be produced, even in a more crowded and warmer world. Yet this hopeful view has to grapple with two likely, connected realities: while population may more than double, production and consumption should more than double.

For two decades, major institutions such as the United Nations (United Nations, Department of International and Economic and Social Affairs, 1992) and the World Bank (Bos et al., 1992) and individual demographers that make 50- to 150-year population forecasts have projected a world population of between eight billion and twelve billion that stabilizes sometime within the next century. Such agreement is qualified by the fact that almost all the forecasters use similar methods and assumptions (Frejka, 1981; Lutz, 1994).

The common and key assumption for long-term forecasts is the completion of the demographic transition, specifically, that at some future date all couples within a country will reduce their births to a level at which they just reproduce themselves and will maintain that level over the next century (see Figure 6). The

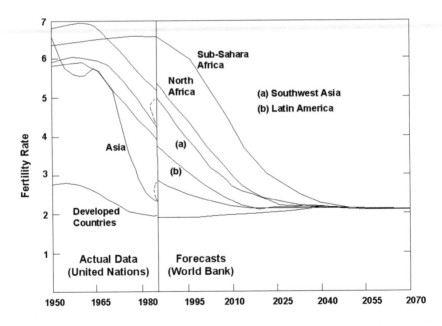

FIGURE 6 Projected fertility rates. NOTE: Different methods explain the present discrepanices in the rates reported by the United Nations and the World Bank. SOURCE: Lee (1991, p. 58).

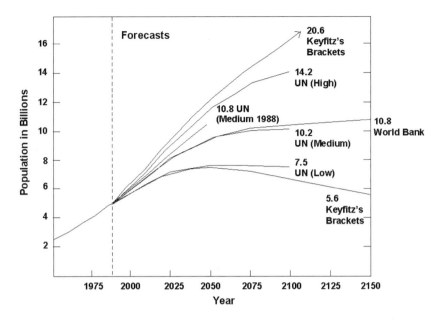

FIGURE 7   World population projections. SOURCE: Lee (1991, p. 59).

dates for when this should happen vary by the forecasters' assessment of the rapidity of the transition. According to a 1990 forecast, for example, it will take place in China by the year 2000, India by 2005, and Nigeria, much later, by 2035. Attaining this level of just reproducing the parents, however, does not mean that the population is stabilized, because the momentum of having a large population of young people just entering their reproductive life pushes up the growth for a long time. Thus, population growth would not diminish to negligible levels until 2075 in China, 2100 in India, and well into the twenty-second century in African nations.

The somewhat arbitrary choice of these dates matters, as do the assumptions about how quickly the death rate declines, and demographers therefore prefer to show low, medium, or high variants of their projections. The current variants of the major projections forecast a medium projection of ten to eleven billion and a low-high range between eight billion and fifteen billion for the end of the next century (see Figure 7). Even this broad range may be too narrow (McNicoll, 1992). Demographers who have attempted to handicap the accuracy of UN forecasts for individual countries made an estimate of the average errors made by their UN brethren. Using these estimates they would even set wider limits, arguing that there is a two to one chance that in the year 2100, global population will fall somewhere between five billion and twenty billion people (Keyfitz, 1981; Lee, 1991; Stoto, 1983).

These ranges assume that errors are equally possible in both directions, but the renewed concern for population is directed toward the upper end. Upward rather than downward creep is suggested by the apparent slowing of the decline in birthrates mentioned above. There are unknowns on the mortality end as well, although simulations of the impact of AIDS, for example, find that despite a death toll in many millions, AIDS has only a small effect on global projections involving billions (Bongaarts, 1996).

Even a doubling of the population could be too much if future consumers use and discard at the levels of Americans, rather than of Africans, today. One study extrapolating "current trends" found that a doubling of population requires a quadrupling of agriculture, a sextupling of energy, and an octupling of the economy if varied and nutritious diets, industrial products, and regular jobs are to be within reach of most of the ten billion people (Anderberg, 1989).

Many find this 2–4–6–8 scenario unbelievable and unsustainable because of the extraordinary increases in production and consumption required by "just" the doubling of the population. Such increases could hardly be accommodated by current technology and practice in a human environment that already has seen substantial transformation of its atmosphere, soils, groundwater, and biota. If environmental catastrophe is to be postponed in such a warmer and more crowded world, it can be done only by maintaining great inequities in human welfare or by achieving different trajectories for technology and development.

As we contemplate what those different trajectories for technology and development might be, we can gauge the outlook through our temporal frames. We appear to be about halfway in numbers into the third great population surge, and the good news from the ages is thus that some relief may lie ahead, albeit in a century or so. Twentieth-century population and consumption growth is totally unprecedented in human history, and the bad news from the millennia is that great civilizations failed to maintain much smaller rates of growth in the past. We also have no news, especially from the centuries: our science can observe but not readily explain past and existing interactions of population, technology, and resources. But, like Malthus, we have theories. To address these interactions; to move beyond theories to practices; to assist in the passage through the Great Climacteric of the next decades—these challenges provide an extraordinary and fulfilling charter for studies of the human environment.

NOTE

1. For the original Latin, see Tertullian, *De anima*, Chapter 30, sentences 3 and 4.

REFERENCES

Anderberg, S. 1989. A conventional wisdom scenario for global population, energy, and agriculture, 1975–2075. Pp. 209–229 in Scenarios of Socioeconomic Development for Studies of Global Environmental Change: A Critical Review, F. L. Toth, E. Hizsnyik, and W. C. Clark, eds. RR-89-4. Laxenburg, Austria: International Institute for Applied Systems Analysis.

Bongaarts, J. 1996. Global trends in AIDS mortality. Population and Development Review 22:21–45.

Bos, E., P. W. Stephens, and M. T. Vu. 1992. World Population Projections. 1992–93 Edition. Baltimore: Johns Hopkins University Press.

Boserup, E. 1965. The Conditions of Agricultural Progress. London: Allen and Unwin.

Boserup, E. 1981. Population and Technological Change: A Study of Long-Term Trends. Chicago: University of Chicago Press.

Bowden, M. J., R. W. Kates, P. A. Kay, W. E. Riebsame, R. A. Warrick, D. L. Johnson, H. A. Gould, and D. Wiener. 1981. The effect of climate fluctuations on human populations: Two hypotheses. Pp. 479–513 in Climate and History: Studies in Past Climates and Their Impact on Man, T. M. L. Wigley and G. Farmer, eds. Cambridge, England: Cambridge University Press.

Brown, H. 1954. The Challenge of Man's Future. New York: Viking Press.

Burton, I., and R. W. Kates. 1986. The great climacteric, 1798–2048: The transition to a just and sustainable human environment. In Geography, Resources, and Environment. Volume II: Themes from the Work of Gilbert F. White; R. W. Kates and I. Burton, eds. Chicago: University of Chicago Press.

Caldwell, J. C. 1994. Fertility in Sub-Saharan Africa: Status and prospects. Population and Development Review 20:179–187.

Caldwell, J., and P. Caldwell. 1987. The cultural context of high fertility in Sub-Saharan Africa. Population and Development Review 13:409–437.

Caldwell, J. C., P. H. Reddy, and P. Caldwell. 1988. The Causes of Demographic Change: Experimental Research in South India. Madison, Wis.: University of Wisconsin Press.

Clark, C. 1968. Population Growth and Land Use. London: Macmillan.

Coale, A. 1973. The demographic transition. Pp. 53–72 in International Population Conference. Vol. 1. Liège: IUSSP.

Cohen, J. E. 1995. How Many People Can the Earth Support? New York: W. W. Norton.

Cohen, M. N. 1977. The Food Crisis in Prehistory. New Haven, Conn.: Yale University Press.

Cohen, M. N. 1990. Prehistoric patterns of hunger. Pp. 56–97 in Hunger in History: Food Shortage, Poverty, and Deprivation, L. F. Newman, W. Crossgrove, R. W. Kates, R. Mathews, and S. Millman, eds. Oxford, England: Basil Blackwell.

Cowan, C. W., and P. J. Watson. 1992. The Origins of Agriculture: An International Perspective. Washington, D.C.: Smithsonian Institution Press.

Davis, K. 1990. Population and resources: Fact and interpretation. Pp. 1–21 in Resources, Environment and Population: Present Knowledge, Future Options, K. Davis and M. S. Bernstam, eds. Supplement to Population and Development Review 16. New York: Oxford University Press.

Deevey, E. 1960. The human population. Scientific American 203:194–204.

Ehrlich, P. R., and A. H. Ehrlich. 1990. The Population Explosion. New York: Simon and Schuster.

Finkle, J. L., and B. C. Crane. 1975. The politics of Bucharest: Population development and the new international economic order. Population and Development Review 1:87–146.

Frejka, T. 1981. World population projections: A concise history. Proceedings of the IUSSP International Population Conference (Manila) 3:505–528.

Glacken, C. 1967. Traces on the Rhodian Shore: Nature and Culture in Western Thought from Ancient Times to the End of the Eighteenth Century. Berkeley, Calif.: University of California Press.

Hernandez, D. J. 1981. The impact of family planning programs on fertility in developing countries: A critical evaluation. Social Science Research 10:32–66.

James. P. 1979. Population Malthus, His Life and Times. London: Routledge and Kegan Paul.

Jevons, W. S. 1865. The Coal Question: An Inquiry Concerning the Progress of the Nation on the Probable Exhaustion of Coal. London: Macmillan.

Johnson, D. L., and H. Gould. 1984. The effect of climate fluctuations on human populations: A case study of Mesopotamian society. Pp. 117–136 in Climate and Development, A. K. Biswas, ed. Dublin: Tycooly International Publishers.

Kates, R. W. 1995. Labnotes from the Jeremiah Experiment: Hope for a sustainable transition. Annals of the Association of American Geographers 85:623–640.

Keyfitz, N. 1981. The limits of population forecasting. Population and Development Review 7:579–593.

Knodel, J., and E. van de Walle. 1976. Lessons from the past: Policy implications of historical fertility studies. Population and Development Review 5:217–245.

Kunitz, S. J. 1986. Mortality since Malthus. Pp. 279–302 in The State of Population Theory: Forward from Malthus, D. Coleman and R. Schofield, eds. Oxford, England: Basil Blackwell.

Lapham, R. J., and W. P. Maudlin. 1984. Family planning program effort and birthrate decline in developing countries. International Family Planning Perspectives 10:109–118.

Lapham, R. J., and W. P. Maudlin. 1987. The effects of family planning on fertility: Research findings. Pp. 647–680 in Organizing for Effective Family Planning Programs, R. J. Lapham and G. B. Simmons, eds. Washington, D.C.: National Academy Press.

Lee, R. 1986. Malthus and Boserup: A dynamic synthesis. In The State of Population Theory: Forward from Malthus, D. Coleman and R. Schofield, eds. Oxford, England: Basil Blackwell.

Lee, R. 1991. Long-run global population forecasts: A critical appraisal. Pp. 44–71 in Resources, Environment and Population: Present Knowledge, Future Options, K. Davis and M. S. Bernstam, eds. Supplement to Population and Development Review 16 (1990). New York: Oxford University Press.

Lutz, W. 1994. The Future Population of the World: What Can We Assume Today? London: Earthscan.

Malthus, T. R. 1798. An Essay on the Principle of Population as it Affects the Future Improvement of Society with Remarks on the Speculations of Mr. Godwin, M. Condorcet, and Other Writers. London: J. Johnson. Reprinted with notes by Bonar, J. 1965. Reprints of Economic Classics. New York: Augustus M. Kelley.

Mathews, R., D. Anderson, R. S. Chen, and T. Webb. 1990. Global climate and the origins of agriculture. Pp. 27–55 in Hunger in History: Food Shortage, Poverty, and Deprivation, L. F. Newman, W. Crossgrove, R. W. Kates, R. Mathews, and S. Millman, eds. Oxford, England: Basil Blackwell.

Maudlin, W. P., and B. Berelson. 1978. Conditions of fertility decline in developing countries, 1965–75. Studies in Family Planning 9:89–147.

McEvedy, C., and R. Jones. 1985. Atlas of World Population History. New York: Penguin.

McKeown, T. 1976. The Modern Rise of Population. New York: Academic Press.

McNicoll, G. 1992. The United Nations' long-range population projections. Population and Development Review 18:333–340.

Meadows, D. H., D. L. Meadows, J. Randers, and W. W. Behrens III. 1972. The Limits to Growth. New York: Universe Books.

Meadows, D. H., D. L. Meadows, and J. Randers. 1992. Beyond the Limits: Confronting Global Collapse, Envisioning a Sustainable Future. Post Mills, Vt.: Chelsea Green.

Peng, P. 1993. Accomplishments of China's family planning program: A statement by a Chinese official. Population and Development Review 19:399–403.

Petersen, W. 1979. Malthus. Cambridge, Mass.: Harvard University Press.

Population Reference Bureau. 1976. World Population Growth and Response: 1965–1975—A Decade of Global Action. Washington, D.C.: Population Reference Bureau.

President's Materials Policy Commission. 1952. Resources for Freedom (Paley Report). Five Volumes. Washington, D.C.: The Commission.

Puitchett, L. H. 1994. Desired fertility and the impact on population policies. Population and Development Review 20:1–55.

Reed, C. A., ed. 1977. Origins of Agriculture. The Hague: Mouton Publishers.

Rindos, D. 1984. The Origins of Agriculture: An Evolutionary Perspective. New York: Academic Press.

Sadik, N. 1991. The State of World Population 1990. New York: United Nations Population Fund.

Sahlins, M. 1972. Stone Age Economics. Chicago: Aldine.

Sherris, J. D. 1985. The impact of family planning programs on fertility. Population Reports. Series J, No. 29. Baltimore: Johns Hopkins University, Population Information Program.

Stoto, M. 1983. The accuracy of population projections. Journal of the American Statistical Association 78(381):13–20.

United Nations. 1993. World Population Prospects: The 1992 Revision. New York: United Nations.

United Nations, Department of Economic and Social Affairs. 1958. The Future Growth of World Population. Population Studies No. 28. New York: United Nations.

United Nations, Department of International Economic and Social Affairs. 1992. Long-Range World Population Projections: Two Centuries of Population Growth 1950–2150. New York: United Nations.

Vogt, W. 1948. Road to Survival. New York: William Sloane.

Walter, J., and R. Schofield. 1989. Famine, disease and crisis mortality in early modern society. Pp. 1–73 in Famine, Disease and the Social Order in Early Modern Society, J. Walter and R. Schofield, eds. Cambridge, England: Cambridge University Press.

Whitmore, T. M., B. L. Turner II, D. L. Johnson, R. W. Kates, and T. R. Gottschang. 1990. Long-term population change. Pp. 25–39 in The Earth as Transformed by Human Action, B. L. Turner II, W. C. Clark, R. W. Kates, J. T. Mathews, and J. Richards, eds. Cambridge, England: Cambridge University Press.

Technological Trajectories and the Human Environment. 1997.
Pp. 56–73. Washington, DC: National Academy Press.

How Much Land Can Ten Billion People Spare for Nature?

PAUL E. WAGGONER

If people keep multiplying and farmers keep farming as they do now, farmers will soon need to grow their crops on twice as large an area as what they use today.[1] Doubling the population without changing the way we farm would expand the cropland from its present tenth of the world's land to about a fifth. More than any other factor, the success farmers have in feeding more people per hectare (ha) will govern what humanity is able to spare for Nature. I capitalize Nature here and throughout to indicate a specific definition, namely, the features and products of the earth itself, as contrasted with those of human civilization.

My essay presumes a population of ten billion people because that seems to be the round number in sight. The billions may level off at ten, or they may grow further (Lutz, 1994; see also Kates, this volume). In either case, we must contemplate ten billion.

I presume also that humanity should spare lots of land for Nature. Proponents of the sparing of land reason about portfolio, money, and ethics. They argue that sparing land for Nature brings security by assuring a portfolio of biological diversity. They assert that Nature saves our money through her free ecosystem services (Norton, 1988). At bottom, however, is the ethical argument that survives quibbling over the utility of genes in a jungle or whether a marsh purifies water more cheaply than does a sewage plant. Although most religions emphasize humanity, even Genesis declares, "Let the waters bring forth swarms of living creatures, and let birds fly above the earth. . . . And God saw that it was good." My title can presume, therefore, that humanity should spare land for Nature without further justification (see Meyer-Abich, this volume).

The following example shows that expecting farmers to spare land is not a

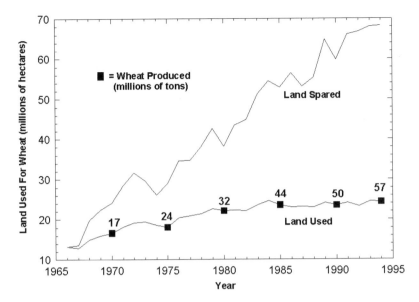

FIGURE 1 The land Indian farmers spared by raising wheat yields. NOTE: The upper line shows the area Indian farmers would have harvested at 1961–1966 yields to grow what they produced. The lower line shows the area they actually harvested. The farmers spared the difference. The numerals attached to the squares show the millions of tons of wheat produced in six exemplary years. SOURCE: Extends a table compiled by Borlaug (1987).

futile wish. From 1961 to 1966, Indian farmers on average grew 0.83 tons of wheat per hectare on 13 million hectares of land. Then, applying the technology of the Green Revolution, they raised production more than fivefold and used only 80 percent more land. Looking back from 1994 to 1961–1966 one can see that Indian farmers spared 44 million ha, about the area of California, by growing more per hectare (see Figure 1). "How much land can ten billion people spare for Nature?" is a farmer's question; asking it is justified, and answering it is not futile.

MAKING DO WITH PRESENT FARMING

The answer to what farmers can grow rests first on what they do grow today, using 11 percent of the world's land, the 1.4 billion ha of cropland. We can translate all agricultural production into food energy, or calories, and protein. Actual national food supplies range from about 1,800 to 3,900 calories and 40 to 130 grams of protein. The US National Research Council recommends between

1,900 and 3,000 calories and 50 to 60 grams of protein per day per person (FAO, 1992; National Research Council, 1989). By dividing today's total calories and protein into rations for ten billion people, we can relate present production to future needs.

As evident in Figure 2, food crops, such as wheat and potatoes, would supply about 1,800 calories; feed crops, such as maize and soybeans, would give another 1,000 calories to each of the ten billion people. Other agricultural products such as tobacco and rubber are neither food nor feed but could be replaced by other crops. This replacement of these other present products would provide little in the form of calories and protein for ten billion people.

Animals appear twice in our accounting. Animal products, mainly meat and eggs, provide a lot of protein and also add some calories. The calories and protein for draft animals require some explanation: they represent consumption by the animals. In 1910 the horses and mules on American farms and in American cities consumed feed that was grown on an area 44 percent as large as that used to cultivate products for domestic use; their replacement by tractors and trucks has been blamed for the American grain surplus of the 1930s (Hassebrook and Hegyes, 1989; US Department of Agriculture, 1962). The present global population of water buffaloes (139 million) and camels (20 million) will surprise a Westerner, as will how much they consume.

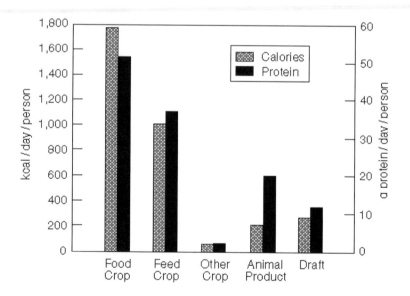

FIGURE 2 World agricultural production of calories and protein and the consumption of draft animals. NOTE: The calories or protein in each class are averages per day per person for ten billion people.

What is the sum of all the categories? If we stopped feeding crops to animals, became vegetarians, and replaced coffee beans with garbanzo beans, the croplands would produce 2,900 consumable calories. What should the animal products add? Even efficient broiler chickens put only about a fifth of the calories they eat into meat.[2] Grazing animals eat more than feed crops, so agriculture, if not cropland, must be credited with some part of the animal calories. Allowing for some further release of calories by reduction in animal numbers but also a continuing role for draft animals, I add 200 calories to the 2,900 calories in crops, bringing the total for ten billion people to 3,100 calories. I simply leave fish on the table without counting its contribution.

The sum of 3,100 calories per day for a population of ten billion exceeds the recommended daily allowances, and it exceeds the 2,920 calories that the Japanese consume today. The same accounting provides an ample amount of protein for ten billion people. This accounting of present farming makes the idea of sustaining a population of ten billion while sparing land for Nature conceivable.

Although today's farming could *sustain* ten billion people, their *wants* are surely more than just sustenance. Prophesying wants is chancy. Because animals eat more calories in feed than they give in milk, meat, or eggs, future wants must encompass *original* calories—those for people plus their beasts. Forecasts of consumption of original calories from rising income alone reach as high as 10,000, but caloric restraint can limit the rise to only 4,400 per person per day (Parikh, 1992; Sanderson, 1988).

Will people who are sticking to their accustomed meaty diets spoil this picture of original calories and my implication that the 3,000 calories supplied by today's agriculture might suffice? Large numbers of people do change what they eat, as widespread rises in meat consumption in conjunction with riches in fact illustrate. Nevertheless, annual American beef consumption peaked in 1976 at close to 60 kilograms per person and has since fallen to about 45 kilograms. The success of McDonald's restaurants, interestingly, has been attributed to potatoes rather than to hamburger, and the consumption of potatoes has accordingly crept up. Since 1910, fat in the American diet has increased 50 percent, but its rise encompassed the opposing trends of much more fat from plants and much less from animals (Kroc, 1977).[3] In summary, by eating a more or less vegetarian diet we can change dramatically the number of people that a plot of land can feed, and over periods of decades large numbers do change diets.

LIMITS TO YIELDS OF FOOD

Suppose we do not simplify our diets and restrain our appetites. Might not global shortages of the essentials needed for photosynthesis still fulfill the Malthusian fears for ten billion or inhibit their ability to spare land for Nature?

At bottom, food comes from photosynthesis, supplied with carbon dioxide (CO_2) and water to combine into carbohydrates, energized by sunlight, and supplemented with fertilizer. Glut has driven fertilizer prices down. Global use has been level since 1988 and in the United States since 1980. Globally, sunlight and, increasingly, CO_2 are also abundant. Because the same pores that admit CO_2 into leaves let water out, an iron correlation attaches photosynthesis to water. But globally the water on land far exceeds the amount needed to grow food for ten billion people.[4]

Although cropland per capita expanded from the year 1700 to 1950, it has since shrunk while per capita food production has risen (Richards, 1990; FAO, 1992).[11] Having techniques that raise yield per hectare and having farmers use these techniques are clearly preeminent in this historic reversal. About 1940, World War II ignited the technological fuel that had been accumulating for a generation in industrial nations, and then, under the banner of the Green Revolution of the 1960s, similar techniques raised yields worldwide. The rising trend, illustrated in Figure 3 by wheat yields in three nations, is familiar.

But when will an upper limit end the trend? The supplies of sunlight and water seem too large to cap yields until well past ten billion. Setting the limit by fitting curves to the actual data in Figure 3 gives too much latitude to pessimism or optimism. I chose, therefore, the real yield grown currently by a contest winner as a prospective limit.

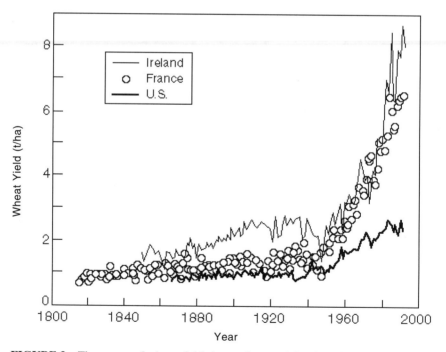

FIGURE 3   The course of wheat yields in tons/hectare (t/ha) in Ireland, France, and the United States.

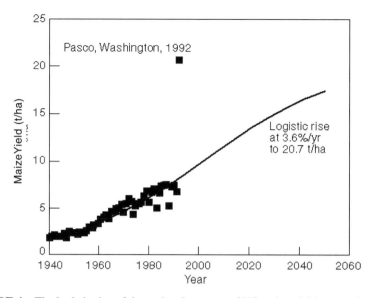

FIGURE 4   The logistic rise of the national average of US maize yields toward a maximum of 21 tons/hectare (t/ha).

Maize, with its efficient photosynthesis, is a productive crop. In 1992 the annual National (US) Corn Growers' Association competition enrolled 2,470 entries from forty-four states (National Corn Growers' Association, 1993, 1994). To enroll, farmers had to enter a minimum of 4 ha of maize and keep accurate production and harvest records. The winning, irrigated field in Pasco, Washington (46° north with a sunny climate) grew a full 21 tons (t) per hectare! There were other yields above 18 t/ha (287 bushels per acre), proving that 21 t was not a fluke. Providing still further proof that it was no fluke, the Pasco farmer came back in abnormally cool, wet 1993 to grow 19.6 t/ha on his supervised area and 16.3 t on 575 ha. Twenty-one tons would feed eighty people 2,900 calories/day for a year. At eighty people per hectare, 125 million ha, or less than a tenth of the present cropland, could support a population of ten billion.

In Figure 4 a logistic curve rises with the actual average American maize yields toward the limit of the winning 1992 yield. Pessimists will worry whether national averages can approach the yield of the irrigated winner. They may reason that yields far above the primitive ones mean more effort must go into maintenance of the yields (Plucknett and Smith, 1986), and they may observe that averages have recently fallen and are further below the trend than typically occurred during 1940–1970. Optimists will trust that new techniques can raise the limit above 21 t and that a relay team of maintenance research and application will steady the annual averages. But surely all will agree that, since a farmer grew a real yield of 21 t on 4 ha while the national average in an industrial nation lies near 7 t, scope remains for raising yields to feed people everywhere while sparing land for Nature.

BUT IN THE END, WILL FARMERS SPARE LAND FOR NATURE?

Production does fluctuate, and people do panic. For example, in the early 1970s fallen production drove food prices up, and US soybean prices doubled. Anxious academics and politicians launched world hunger studies. Then production recovered, sinking prices and bankrupting farmers. Looking beyond fluctuations—and grain prices in 1996 show we are suffering one right now—takes steady nerves.

The logistic curve extending past improvements in yields toward 21 t/ha could mislead humanity into thinking that an unseen hand lifts yields effortlessly. In fact, vigorous research and enterprising farmers do the lifting. Remembering the lag of decades between discoveries and their impact on world averages, one asks whether any innovations are on the shelf that can raise yields soon. Heralded for decades, some techniques from biotechnology now sit prominently poised for application, and both scientists and practical people expect them to raise yields as well as protect crops and lessen environmental harm (US Congress, Office of Technology Assessment, 1992; Weiss and Brayman, 1992).

Concrete, statistical evidence that techniques remain to be more fully used appears in the comparison of best and average practices of maize farmers (Iowa Crop Improvement Association, personal communication; see also various years of the *FAO Yearbook* and the US Department of Agriculture's *Agricultural Statistics*). Figure 5 displays the trends since 1960 of maize yields grown by the winners of the Iowa Master Corn Growers' Contest and also the trends of average yields by Iowa and world farmers. In percentages, the annual gains by world and Iowa averages do exceed the gain by Iowa Masters. Absolutely, however, the Masters gained 0.14 t/ha annually, more than the Iowa average and twice the gain of the world average. The winners of the Iowa Master Soybean Growers' Contest also steadily stay ahead of the Iowa average soybean yields. The reality of winners staying steadily ahead of averages confirms that new technology remains at hand for American farmers. A survey of irrigated Pakistani farms shows a similar gap between master and average yields. Ahmad's tabulation of yields of major crops showed that progressive Pakistanis grow about three times the average yields (Ahmad, 1987). Technology remains on a nearby shelf for farmers everywhere.

Because widespread use lags behind discovery by decades, the inventory on the shelf cannot be filled on need but must be replenished continually. The expenditures by the Consultative Group on International Agricultural Research provide an index of effort to refresh the inventory worldwide. Expenditures, which are on the order of a quarter billion dollars, peaked in 1989 and in 1994

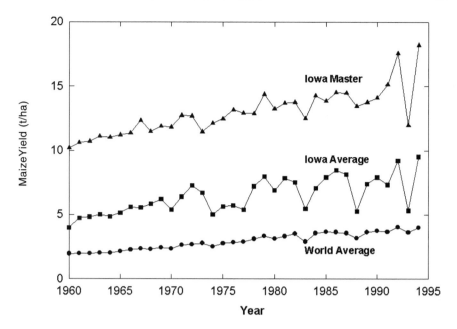

FIGURE 5 The trends since 1960 of maize yields in tons/hectare (t/ha) grown by the winners of the Iowa Master Corn Growers' Contest and also of average yields of Iowa and world farmers. NOTE: The rising trend of per-year yields for Iowa Masters is 1.1 percent, or 0.14 t/ha; for Iowa, the average is 1.5 percent, or 0.10 t/ha; and for world maize growers, the average is 2.2 percent, or 0.06 t/ha.

were about 20 percent below the peak.[5] So while the shelf currently holds technology, what it will hold in a few decades causes us to worry justifiably.

Technology left on the shelf butters no parsnips. Whether it will be employed depends on the profit the farmer foresees or the rules that discourage him. In *Transforming Traditional Agriculture,* Schultz (1964) argues that even poor farmers in poor places do profitable things. A book with the illuminating title of *The Bias Against Agriculture* (Bautista and Valdes, 1993), however, tells how societies have both discouraged and encouraged farmers' production. For example, in Peru during 1969–1973 favors for industry and price controls on farm products lowered the production of farm products. In Zaire during 1966–1982 price controls on food to depress real farm wages, as well as taxes on farm exports to provide cheap credit for industry, were designed to encourage industry; they cut the growth of food production in half and of export crops by even more. None of these workings of an invisible hand would have surprised Adam Smith.

An invisible hand also induces people and institutions to invent and apply technology. The ratio of fertilizer to land prices induced about the same applications of fertilizer in countries as unlike as Japan and the United States. Passing time changed the output per worker and per unit of land similarly in different countries (Hayami and Ruttan, 1985). By incentives and rules, nations will replenish the technology on the shelf and lead farmers to use it or not, sparing land for Nature or not.

Nations could choose Draconian rules against expanding cultivation and favoring intensive farming to spare more land for Nature. But they must beware the price of food. Mobs have taught the rulers of Rome, revolutionary France, and modern states that costly bread incites riots. Thomas Malthus foresaw that no sensible politician would do away with farm animals and require people to eat only potatoes.

If one includes improved technology in an analysis, the desired outcome can be envisioned without exploding prices. The outcome requires a per hectare productivity rise of 2 percent annually; this target exceeds recent increases and projected percentage rises for US crops but not the rise of global maize yield or of US land productivity from 1950 to 1979. A reasonable analysis can produce an annual decline of food prices of 0.5 percent, which matches the 1900–1984 fall of world prices of the main agricultural products.[6]

DOES WATER CLOUD THE VISION?

Despite the abundance of water overall, its uneven distribution among regions and its capricious variation among seasons plague farming. The brute expansion of irrigation grows harder. Nevertheless, opportunities to grow more crops with the same amount of water kindle our hope. People usually see the last oasis and pin their hopes on engineering—lining, metering and timing, trickle, surge, and drip (Postel, 1992).

A peculiarity about evaporation creates a paradoxical but even greater opportunity: Bumper crops consume only a little more water than do sparse ones. Doubling yield doubles water-use efficiency, as we see in Figure 6. Consider irrigation with 450 mm of water. In a survey of Pakistani farms, increasing fertilizer from 20 kg/ha to 100 kg/ha raised yield by 40 percent. Because no more water was used, fertilizer also raised water-use efficiency by 40 percent. Ahmad wrote, "Water cannot be considered to have become a real constraint to meeting the world food supplies as long as there is the scope for manipulation of the various underlying factors for . . . increasing . . . production" (Ahmad, 1987).

Another paradoxical opportunity to make water go further is to supplement rain. Water that supplements rain supplies the fast evapotranspiration that raises water-use efficiency. For example, the water-use efficiency of sorghum in Texas doubled when the water supply raised evapotranspiration from 250 mm to more than 700 mm; water similarly raised the efficiency of maize in five US states (Jensen, 1984) The simplest rationale for irrigation in humid places is that rain provides some of the needed water for free.

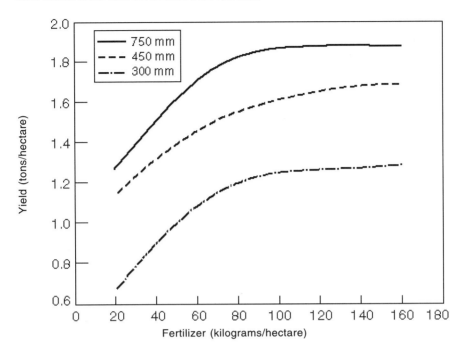

FIGURE 6 The complementarity of increasing irrigation and fertilizer. NOTE: The three curves show different amounts of applied water, measured as precipitation in millimeters (mm).

HOW MUCH WILL HIGH YIELDS TARNISH THE LAND?

If farmers increase their yields with techniques that harm the surroundings, they will spare land, but the external effects may tarnish their victories.

Farmers do many things on each area of land they crop. In general, higher yields require little more clearing, tilling, and cultivating than lower yields. Protecting a plot of lush foliage from insects or disease requires only a little more pesticide than does sparse foliage. Keeping weeds from growing in deep shade beneath a bumper crop may require less herbicide per field than keeping them from growing in thin shade. The amount of water consumed is more or less the same per area whether the crop is abundant or sparse, and growing higher yields distills away only a little more water and leaves only a little more residue of salt than lower yields.

Seed is planted per plot; choosing a higher yielding variety does not affect the surroundings. If the improved variety resists pests, it lessens the external effect of pesticides compared to a sprayed crop. If the pests in a crop had gone uncontrolled and had decreased the yield, the new variety and its higher yield would be free, environmentally. By minimally changing the external effects of things that farmers do per area, lifting yields will thus lower the effects per yield.

On the other hand, farmers use more of some things to raise the yield of their crops. For example, farmers apply more fertilizer per plot to raise yields. Does this leak more fertilizer into the surroundings per yield?

Consider again the complementarity of water and fertilizer (Figure 6). A given yield requires more fertilizer with 300 mm than 750 mm of irrigation. Consider the yield of 1.4 t/ha. Even an infinite amount of fertilizer and great fallout into the surroundings would not produce 1.4 tons on the field irrigated with 300 mm of water. But 50 units of fertilizer would grow 1.4 tons on the field irrigated with 450 mm of water. The hectare irrigated with 750 mm of water would get the same 1.4 tons from only 30 units and would therefore have less environmental fallout.

For a given yield, optimum conditions for growth and high yield lessen the fallout of such things as silt, pesticides, and fertilizer into the surroundings. If factors that must be increased per plot to raise the yield are improved in step, their improved coordination may diminish the fallout.[7]

STRAWS IN THE WIND

Having reviewed some of the elements of farming, we must look at how global cropland and production are actually changing at the macro level. From 1975 to 1990 cropland expanded by only 3 percent, but from 1969–1970 to 1988–1990 the supply of calories per capita rose by 11 percent (FAO, 1992). Because of the rising yields, farmers grow surpluses today, driving prices down.

To combat the bankruptcy of farmers, prices are supported, and farmers are given incentives to idle their cropland. So far in the 1990s about a fifth of US cropland has typically been idled by government programs.[8] The geographers Deborah and Frank Popper have made vivid the reversion of farms to range with their phrase "The Buffalo Commons" (Matthews, 1992). Looking forward, the Dutch projected changes from the present farmland in nations of the European Union to the year 2015 (Rabbinge et al., 1992). Diverse scenarios built around liberal trade, employment policies, and environmental regulation all shrank farm-land by 40 percent or more. Straws in the wind hint that land can be spared.

A SCENARIO FOR SUCCESS

A tally of strategies to lessen deforestation is a good place to start the search for a scenario about sparing land for Nature. A strategy of economic development to attract settlers away from treasured forests takes too long. Encouraging migra-tion to places other than the lands we wish to protect is usually insufficient to deflect immigrants. In the minds and meeting rooms of environmentalists, desig-nating Nature reserves may stop hungry people from clearing plots; but this is not the case outdoors. And reserves for extractive but sustainable forestry support few people. On the other hand, eliminating the need to abandon land that is already cleared by maintaining or restoring productivity offers some hope. Experiments for eight years on soil representing the Amazon basin grew undiminished yields of about 7 t/ha; this productivity has continued over seventeen years for forty crops (Sanchez et al., 1982, 1990; World Bank, 1992).

Because numbers can impart a misleading aura of accuracy, I have written more about directions than precise numbers. In the end, however, the question "How much?" calls for numerical answers and familiar images of space such as India or Amazonia. The plot for my quantitative scenario relates the area potentially spared to: *1)* A reference area, which I shall set at 2.8 billion ha of cropland. This is twice the size of the present cropland, six-tenths of the present cropland plus permanent pasture, and a fifth of the land in the world. If farmers use less than 2.8 billion ha as the population multiplies from about five to ten billion, I assert that they spare land for Nature. *2)* Diet, with a daily use of calories from agricultural products varying from about 3,000 to 6,000 calories per capita. *3)* Yield, which can vary from 4 to nearly 80 million calories (Mcal) per ha.

Some examples of yields in tons (t) and corresponding Mcal/ha are: wheat in an arid African nation, 1 t and 4 Mcal/ha; wheat in North America, 3 t and 12 Mcal/ha; wheat in Europe, 6 t and 24 Mcal/ha; wheat in Ireland or maize in the United States, 9 t and 35 Mcal/ha; potatoes in Maine or Ireland, 30 t and 18 Mcal/ha; and maize from the field of the national winner in Pasco, Washington, 20 t and 78 Mcal/ha. The 12 quadrillion calories produced by agriculture plus consumption by draft animals, which is shown in Figure 2, divided by the world's 1.4 billion ha of cropland produces an average of about 8.5 Mcal/ha or 2 t/ha.

To support ten billion people consuming 3,000 cal/day, farmers averaging the yield of wheat in arid Africa would spare none of the 2.8 billion ha of the reference area (Figure 7). If the ten billion consumed 6,000 cal/day, the yield of 4 Mcal/ha would spread over an additional 2.8 billion ha of other land. On the other hand, an average yield of 16 Mcal/ha, one-third less than present European wheat, would spare much of the land. Averaging 16 Mcal/ha, farmers would be able to support ten billion people consuming 6,000 cal/day on the present cropland, sparing half of the reference area. If the ten billion consumed only 3,000 cal/day, 16 Mcal/ha would spare for Nature about 600 million hectares of present cropland, the area of the Amazon basin. Above 24 Mcal or 6 t per ha, farmers will use little cropland, globally sparing an area of today's cropped hectares equal to the land of India, even when people consume 6,000 cal/day. If during the next sixty to seventy years the world farmer reaches the average yield of today's US corn grower, the ten billion will need only half of today's cropland while they eat today's American calories.

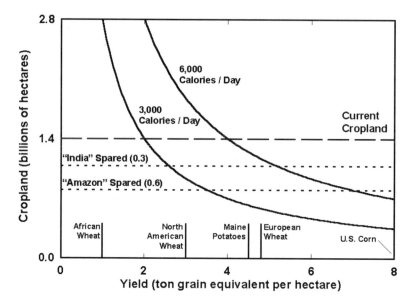

FIGURE 7. The sparing for Nature of a reference area of 2.8 billion hectares of cropland by farmers raising yields for ten billion people consuming 3,000 or 6,000 calories daily.

SURPRISES, BAD AND GOOD

Orderly people instinctively turn to experts for projections that will protect them from surprises. Unfortunately, a century of scientific bloopers by heavy-weights beginning with Lord Kelvin disabuses their instinct.[9] Looking ahead to what ten billion people might be able to save for Nature, I could be daunted by the experts' historic lack of foresight. I could play it safe by conceiving a list of surprises and writing that "all these might happen." Alas, the forecaster who plays it safe by ending all predictions with "but it may snow" is worthless. Beyond listing surprises that could happen, I must suggest which of them are likely to occur and admit that good as well as bad surprises may be in store. From this list of conceivable surprises, I choose four likely ones: fewer than ten billion people, climate change, new pests, and new breakthroughs.

Because growing income and social security have been credited with the slowing of population growth, I look at the speedy economic growth of China and other Asian nations and wonder whether it might check the multiplication of their great populations. The Black Death of the fourteenth century left Europe too small for its clothes, and in 1918 the influenza pandemic left twenty million dead in just a few months. Thus, a surprise like wealth or a pandemic could slow population growth. Then, more land would be spared for Nature in the twenty-first century as it was in the fourteenth.

Heralded for more than a decade, climate change may not come as a surprise. But just as some unexpected happening is not necessarily a surprise while its specific quality is, so it is with climate change. During a debate about supersonic airplanes, cooling associated with their emissions was projected to have a dire impact. Then, lengthening observations of rising CO_2 levels brought projections of warming and drying, with more projections of dire impacts. When the American breadbasket turned dry in 1988, the warmer, drier climate seemed at hand. But during 1993, floods in the American heartland discounted predictions made only five years earlier. Computer simulations, of course, had disagreed all along about whether rising CO_2 would make North America drier or wetter. I place climate change among the surprises.

If cropland in temperate climates becomes hot or dry, yields will fall, and land may be taken from Nature to be used for crops. On the other hand, if cropland that is too cold warms and that which is too dry moistens, yields will rise, saving other land for Nature. Conflicting and changing projections and experience mean we are wise to diversify our portfolios in anticipation of the surprises (CAST, 1992).

Weekly forecasts of pest infestation and the weather affecting them underpin modern pest management. The record shows, however, that pests are shifty, and they may cause disastrous outbreaks and epidemics. New, surprising fungi caused both the Irish potato famine of the 1840s and the Southern corn leaf blight of 1970. "History warns that new pests will appear but provides no data for a model that tells where and when newcomers will appear or what they will be like. The required warning system of sharp, exploring eyes in the field is old-fashioned but remains our most effective approach" (National Research Council, 1976). Surprising pests could lessen the sparing of cropland for Nature.

Scientific breakthroughs, or their failure to appear, could also violate our plans and expectations. In Figure 4 yields rise at a declining rate toward a ceiling that is set by crops that are already grown. This projection assumes that societies will continue to encourage, scientists will continue to discover, and farmers will continue to venture toward that ceiling. Disorder from Dushanbe and Sri Lanka to Kigali and Sarajevo to Port-au-Prince and Monrovia renders hopeless encouragement by some nations. A decrease in money implies a decline in agricultural research. So, a surprise could arrest the trend that is shown in Figure 4.

Yet I would like to point out that some surprises are happy ones. It is optimistic but still rational to hope that some breakthroughs will become practice before the population reaches the ten billion mark, surprising me as rising yields after 1900 would have surprised Malthus and even a writer at the turn of the century. The distance between average yields and the actual (not theoretical) 20 tons of grain on the hectares in Pasco, Washington, provides room for a surprise. The surprise of leaps in productivity and new forms of food production would likely dislocate farming and displace farmers as changes have rapidly and cruelly done since 1940. But it would spare land for Nature.

IN THE END

If people keep eating and multiplying and farmers keep tilling and harvesting as they do today, the imperative of food will take another tenth of the land away from Nature. So farmers work at the hub of sparing land for Nature.

By eating different species of crops and a more or less vegetarian diet people can change the number that a plot can feed. And large numbers of people do change their diets. The calories and protein available from present cropland could provide a vegetarian diet to ten billion people. A diet requiring food and feed totaling 6,000 calories daily for ten billion people, however, would overwhelm the capability of present agriculture on present cropland.

The global totals of sun, CO_2, fertilizer, and even water could produce far more food than what ten billion people need. Encouraged by incentives, farmers combine natural resources with new technology to raise more crop on each plot, keeping food prices down despite the rising population. Differences in yields among nations and between average and best performance continue to show that yields can be raised much more.

For each ton of production, growing more food per plot lessens the fallout of such things as silt and pesticides into the surroundings. If factors such as water and fertilizer are improved in step, fallout may be diminished. Although the uneven distribution of water among regions and its capricious variation among seasons plague farming, opportunities to raise more crops with the same volume of water kindle our hopes for the spread of high yields.

Rising yields have shrunk European and American cropland for decades, and governments pay farmers to keep fields idle. Globally, cropland has been roughly level since the middle of the twentieth century. If average fields in the world sixty or seventy years hence, when we are likely to number ten billion, yield as much food as today's potato fields in Ireland, wheat fields in France, or corn fields in Iowa, large portions of the land currently in crops can revert to Nature. This will not happen by itself, nor will it happen if today's scarcity of grain transfixes us. Countering humanity's multiplying population and wealth to spare habitat for Nature requires never-ending research, encouraging incentives, and smart farmers.

NOTES

1. This article briefly answers the question in the title. An ample answer with full citation of its foundation has been published as Task Force Report No. 121, by the Council for Agricultural Science and Technology (see CAST, 1994). Fallout into the environment is more fully examined in Waggoner (in press).

2. Calculated at 2.6 times the weight of feed, containing 4,000 cal/kg, to produce 1 times the weight of meat, containing 2,200 cal/kg. Feed per meat from US Department of Agriculture (1992).

3. See also various years of the US Department of Agriculture's *Agricultural Statistics*. The book for 1972 tabulates the annual diet from 1909 to 1970.

4. "The staggering conclusion. . . is that 1,000 billion people could live from the earth if photosynthesis is the limiting factor!" (deWit, 1967). Fallen fertilizer prices are reported, e.g., in Freeport-McMoRan (1993). Photosynthesizing 1 to 6 grams of biomass consumes a thousand grams of water. At this ratio, the global evaporation from land matches photosynthesis to feed 400 billion. For details see Waggoner (1994).

5. Expenditures and budgets for 1972–1994 furnished by Ralph Cummings Jr., US AID, and adjusted to 1983 dollars by the US consumer price index. See also Abelson (1995).

6. For past food prices see Binswanger et al. (1985). The analysis, which W. D. Nordhaus devised, is reported fully in Waggoner (1994).

7. The annunciation of the principle of a factor only raising yield when other factors are not limiting coincides with the origins of plant physiology and agronomy. Recently, it has been related to the environmental effects of intensive farming, as in deWit (1992).

8. The Food and Agricultural Policy Research Institute (FAPRI) tabulated the area planted to fifteen principal crops in the United States and the area idled by two programs identified by the acronyms ARP/PLD/0-92 and CRP. The idled areas have been, or are projected to be, steady from 1989 to 1997, but FAPRI projects them to decline after 1997 (FAPRI, 1992).

9. The title of a book shows the feet of clay: The *Experts Speak: The Definitive Compendium of Authoritative Misinformation* (Cerf and Navasky, 1984). When the US president appointed distinguished scientists and engineers to report on technology that would matter to the nation during coming decades, they missed antibiotics, radar, space exploration, and jet-engine aircraft. "In fact, if you were to ask what were the exciting things that happened over the next several decades, they missed all of them, every one" (Townes, 1991).

REFERENCES

Abelson, P. H. 1995. International agriculture. Science 268:11.

Ahmad, M. 1987. Water as a constraint to world food supplies. Pp. 23–27 in Water and Water Policy in World Food Supplies, W. R. Jordan, ed. College Station, Tex.: Texas A&M University Press.

Bautista, R. M., and A. Valdes, eds. 1993. The Bias Against Agriculture. San Francisco: Contemporary Studies Press.

Binswanger, H. P., Y. Mundlak, M.-C. Yang, and A. Bowers. 1985. Estimation of Aggregate Agricultural Response. Agricultural Research Unit Report 48. Washington, D.C.: World Bank.

Borlaug, N. E. 1987. Making institutions work—a scientist's viewpoint. Pp. 387–395 in Water and Water Policy in World Food Supplies, W. R. Jordan, ed. College Station, Tex.: A&M University Press.

CAST (Council for Agricultural Science and Technology). 1992. Preparing US Agriculture for Global Climate Change. Report 119. Ames, Iowa: CAST.

CAST (Council for Agricultural Science and Technology). 1994. Task Force Report No. 121. Ames, Iowa: CAST.

Cerf, C., and V. Navasky. 1984. The Experts Speak: The Definitive Compendium of Authoritative Misinformation. New York: Pantheon.

deWit, C. T. 1967. Photosynthesis: Its relationship to overpopulation. Pp. 315–320 in Harvesting the Sun, A. San Pietro, F. A. Greer, and T. J. Army, eds. New York: Academic Press.

deWit, C. T. 1992. Resource use in agriculture. Agricultural Systems 40:125–151.

FAO (Food and Agriculture Organization). 1992. FAO Yearbook 1991. Vol. 45. Rome: FAO.

FAPRI (Food and Agriculture Policy Research Institute). 1992. P. 83 in 1992 US Agricultural Outlook. Staff Report 1-92. Ames, Iowa, and Columbia, Mo.: FAPRI.

Freeport-McMoRan. 1993. 1992 Annual Report. New Orleans: Freeport-McMoRan.

Hassebrook, C., and G. Hegyes. 1989. Choices for the Heartland. Walthill, Nebr.: Center for Rural Affairs.

Hayami, Y., and V. W. Ruttan. 1985. Agricultural Development. Baltimore: Johns Hopkins University Press.

Jensen, M. E. 1984. Water resource technology and management. Pp. 142–166 in Future Agricultural Technology and Resource Conservation, B. C. Barton, J. A. Maetzold, B. R. Holding, and E. O. Heady, eds. Ames, Iowa: Iowa State University Press.

Kroc, R. 1977. Grinding It Out: The Making of McDonald's. Chicago: Henry Regnery.

Lutz, W. 1994. The Future Population of the Earth: What Can We Assume Today? London: Earthscan.

Matthews, A. 1992. Where the Buffalo Roam. New York: Grove Weidenfeld.

National Corn Growers' Association. 1993 and 1994. Tabulations of the Contestants in the 1992 and 1993 Maize Yield Contest. St. Louis: National Corn Growers' Association.

National Research Council. 1976. P. 128 in Pest Management in Climate and Food. Washington, D.C.: National Academy of Sciences.

National Research Council, Subcommittee on the Tenth Edition of the RDAs. 1989. Pp. 33 and 285 in Recommended Dietary Allowances. Tenth Revised Edition. Washington, D.C.: National Academy Press.

Norton, B. 1988. Commodity, amenity, and morality: The limits of quantification in valuing biodiversity. Pp. 200–205 in Biodiversity, E. O. Wilson, ed. Washington, D.C.: National Academy Press.

Parikh, K. S. 1992. Agricultural and food system scenarios for the 21st century. In Agriculture, Environment and Health: Toward Sustaining Development in the 21st Century, V. W. Ruttan, ed. Minneapolis: University of Minnesota Press.

Plucknett, D. L., and N. J. H. Smith. 1986. Sustaining agricultural yields: As productivity rises, maintenance research is needed to uphold the gains. BioScience 36:40–45.

Postel, S. 1992. Last Oasis. New York: W. W. Norton.

Rabbinge, R., et al. 1992. Ground for Choices. Reports to the Government. Vol. 42. The Hague: Netherlands Science Council for Government Policy.

Richards, J. F. 1990. Land transformation. In The Earth as Transformed by Human Action, B. L. Turner et al., eds. Cambridge, England: Cambridge University Press.

Sanchez, P., D. E. Bandy, J. H. Villachica, and J. J. Nicholaides. 1982. Amazon Basin soils: Management for continuous crop production. Science 216:821–827.

Sanchez, P., C. A. Palm, and T. J. Smyth. 1990. Approaches to mitigate tropical deforestation by sustainable soil management practices. In Soils on a Warmer Earth, H. W. Scharpenseel, M. Schomarker, and A. Ayoub, eds. Developments in Soil Science 20:213.

Sanderson, F. H. 1988. The agro-food filiere: A macroeconomic study on the evolution of the demand structure and induced changes in the destination of agricultural outputs. In The Agro-Technological System Towards 2000, G. Antonelli and A. Quadrio-Curzio, eds. North Holland: Elsevier Science Publishers.

Schultz, T. W. 1964. Transforming Traditional Agriculture. New Haven, Conn.: Yale University Press.

Townes, C. H. 1991. P. 17 in Cosmos Club Bulletin, October. Washington, D.C.: Cosmos Club.

US Congress, Office of Technology Assessment. 1992. Pp. 133–138 in A New Technological Era for American Agriculture. OTA-F-474. Washington, D.C.: Office of Technology Assessment.

US Department of Agriculture. 1962. Agricultural Statistics. Washington, D.C.: US Government Printing Office.

US Department of Agriculture. 1992. Agricultural Statistics. Washington, D.C.: US Government Printing Office.

Waggoner, P. E. 1994. How Much Land Can Ten Billion People Spare for Nature? Task Force Report, No. 121. Ames, Iowa: Council for Agricultural Science and Technology.

Waggoner, P. E. In press. How much more land can American farmers spare? In RCA III Symposium on Crop and Livestock Technologies: Proceedings, B. C. English, R. L. White, and L.-H. Chuang, eds. Washington, D.C.: National Resource Conservation Service, US Department of Agriculture.

Weiss, C., and S. E. Brayman. 1992. Assessment of biotechnology: The 'gene' revolution. Pp. 1–7 in Biotechnology and Development: Expanding the Capacity to Produce Food. New York: United Nations.

World Bank. 1992. Development and the environment: World development indicators. World Development Report 1992. Oxford, England: Oxford University Press.

Technological Trajectories and the Human Environment. 1997.
Pp. 74–88. Washington, DC: National Academy Press.

Freeing Energy from Carbon

NEBOJŠA NAKIĆENOVIĆ

The doing of more with less attests to the practical advancement of societies. In fact, labor, capital, and inputs of other factors to the economy have demonstrably decreased per unit of output and value added since the beginnings of the industrial revolution some two hundred years ago. These increases in the productivity of resources owe to numerous technical and organizational innovations and to an enormous accumulation of knowledge and experience.

A portion of the increases in productivity is attributable simply to the increasing scale of activities, also made possible by technical and organizational innovations. Often with greater size, cost decreases and efficiency increases within specific frames. For example, in building electricity-generating plants a long-standing rule of thumb was that the cost of the plant would grow with two-thirds the power of its size. We are uncertain now where we stand with respect to optimal scale of many facilities and systems, but it seems likely that considerable opportunities to lift efficiency remain.

Perhaps more important than simply size and more certain to continue yielding productivity gains is the accumulation of knowledge and experience. Growth in output in an economic system with suitable incentives tends to bring positive returns of its own. This process is sometimes referred to as "learning by doing." Analysis of learning curves in a range of industries, beginning with the manufacture of aircraft, has provided ample evidence that the costs per unit of output decrease rapidly at a rate proportional to the doubling of the output (Argote and Epple, 1990; Christianson, 1995).

Energy industries and energy systems are not exceptional. This essay will demonstrate that large secular decreases in energy requirements per unit of eco-

74

nomic output have been achieved throughout the world, as we have learned better how to make, operate, and use energy systems. Furthermore, the emissions of carbon dioxide from energy systems, coming from the combustion of the carbon molecules that wood, coal, oil, and gas all contain, have also decreased per unit of energy consumed. This *decarbonization* of the energy system proves to be emblematic of its entire evolution.

At the same time, because of population and general economic growth, absolute world consumption of energy (and many other resources) has increased, especially in the more industrialized countries. This absolute growth often dominates environmental news and views. Rising carbon dioxide emissions are the main contributor to fears of global climatic change. This and other environmental concerns associated with carbon makes energy free from carbon a highly desirable goal for the energy system. The fact that energy and most of the other factor inputs have decreased per unit of output over long periods of time provides a fresh basis on which to project the range of possible future resource use and emissions.

A glance at the changes in labor and materials requirements helps to establish the context and the pervasiveness of the phenomenon that we will observe most closely in energy. Since 1860, the number of hours that workers in the industrialized countries are engaged in paid work each year has generally decreased by half (Figure 1). Though the Japanese bucked the trend for several decades around mid-century and continue to work more than their European and American counterparts, they too are working less. Taking into account the dramatic increase in individual income and consumption over the period, we know that the labor requirements per unit of income and output decreased much faster than the number of hours worked. Furthermore, because life expec-tancy increased by several decades during this period, the years of paid work required to sustain lifelong consumption for a worker at prevailing levels decreased from about three-quarters of a lifetime to less than one-half (Ausubel and Grübler, 1995).

Decreases in requirements for many materials are similarly dramatic (see Wernick et al., this volume). For example, in the United States, which is quite representative of industrialized countries in this regard, steel use declined from about 70 kilograms per $1,000 of GNP (in 1983 dollars) in 1920 to about one-third that level in recent years; cement per GNP in the United States has dropped by about half since 1960 (Williams et al., 1987). However, this *dematerialization* of the economy is varied. In some cases, a lighter steel beam does the work of an earlier, heavier one. In other cases, new materials replace the steel. In contrast, demand per GNP has grown steeply since mid-century for certain petrochemicals (such as ethylene) and for advanced composite materials. Requirements for paper per GNP have been rather flat since about 1930.

Analysis of energy materials and decarbonization may in practice shed light on the question of dematerialization. Because energy is one of the most important

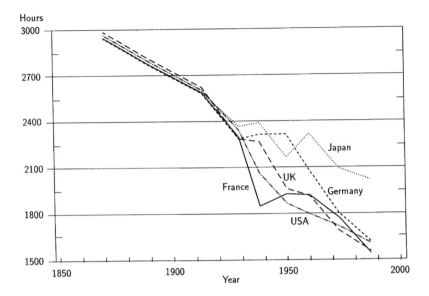

FIGURE 1   Annual working hours in five industrialized countries from 1860 to 1990, expressed in total working hours per year. NOTE: Hours spent on sick leave, strikes, and holidays are subtracted from the formal working time. SOURCE: Ausubel and Grübler (1995) and Maddison (1991).

factor inputs and is embedded in most materials, products, and services, decreases in specific energy requirements can also decrease the intensity of materials use. The carbon content of energy and the subsequent carbon dioxide emissions form the largest single mass flow associated with human activities, excepting water. Current annual global carbon emissions are about 6 billion tons, or more than 1,000 kilograms per person on the planet. In comparison, the global steel industry annually produces about 700 million tons, or about 120 kilograms per person. Therefore, decarbonization can contribute in a large way to dematerialization.

Let us now turn to energy and examine the savings of carbon that have been obtained, why they may have occurred, and whether future savings may be sufficient to spare the environment some unwanted heat.

THE GLOBAL HISTORY OF ENERGY AND CARBON SAVINGS

To form a picture of carbon use, we need to be able to sum and compare its appearances. One way is to index carbon by the ratio of carbon atoms to hydrogen atoms in the energy sources that contain both of these fuels. Fuelwood has the highest effective carbon content, with about ten carbon atoms per hydrogen atom. If consumed without a compensating growth of biomass, which occurred in the past and still occurs in most developing countries, fuelwood thus produces higher carbon emissions than any of the fossil energy forms. Among fossil energy sources, coal has the highest carbon-to-hydrogen ratio, roughly one to one. Oil has on average one carbon for every two hydrogen atoms, and natural gas, or methane, has a ratio of one to four. Using these types of elemental analyses, we can estimate the total amount of carbon contained in a given supply of an individual fuel or a mix of fuels and compare this amount to energy consumed or associated economic output.

Decarbonization can then be expressed as a product of two factors: *1)* carbon emissions per unit of energy consumption; and *2)* energy requirements per unit of value added, which is often called energy intensity. Available data allow us to assess with reasonable confidence the trend for each of these factors since the nineteenth century for major energy-consuming regions and countries, such as the United States and the United Kingdom, and thus for the world as a whole as well. As Figure 2 shows, the ratio of carbon emissions per unit of primary energy consumed globally has fallen by about 0.3 percent per year since 1860. The ratio has decreased because high-carbon fuels, such as wood and coal, have been continuously replaced by those with lower carbon content, such as gas, and also in recent decades by nuclear energy from uranium and hydropower, which contain no carbon.

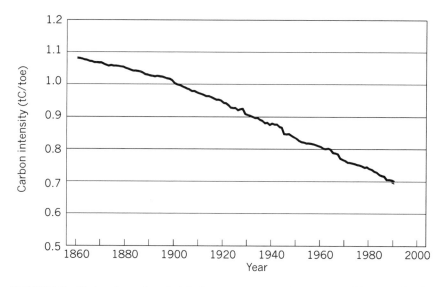

FIGURE 2 Carbon intensity of global energy consumption, expressed in tons of carbon per ton of oil equivalent energy (tC/toe).

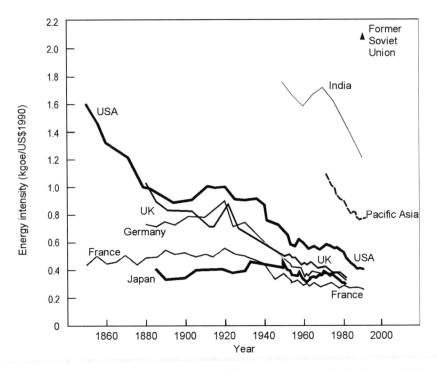

FIGURE 3   Primary energy intensity, including biomass, per unit of value added from 1855 to 1990, expressed in kilograms of oil equivalent energy per GDP in constant 1990 US dollars (kgoe/US$1990).

The historical rate of decrease in energy intensity per unit of value appears to have averaged about 1 percent per year since the mid-nineteenth century and about 2 percent per year in some countries since the 1970s. The overall tendency is toward lower energy intensities, although paths of energy development in different countries have varied enormously and rather consistently over long periods (Figure 3). For example, France and Japan have always used energy more sparingly than the United States, the United Kingdom, or Germany. In some of the rapidly industrializing countries, such as China or Nigeria, *commercial* energy intensity is still increasing. Because commercial energy replaces traditional energy forms not sold in the markets whose transactions find their way into national statistical data, total energy intensity may diminish while commercial energy intensity increases. The present energy intensity of Thailand resembles the situation in the United States in the late 1940s. The energy intensity of India and its present improvement rates are similar to those of the United States about a century ago.

    Combining the two factors of carbon intensity and energy intensity (Figure

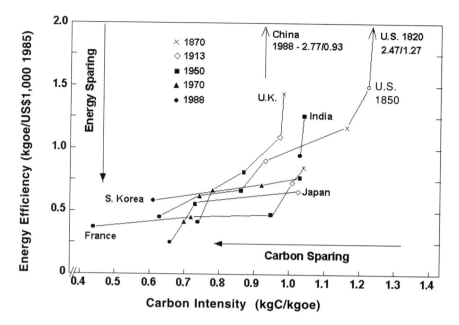

FIGURE 4 Global decarbonization by carbon and energy sparing from 1870 to 1988, expressed in kilograms of carbon per kilogram of oil equivalent energy (kgC/kgoe) and in kilograms of oil equivalent energy per $1,000 of GDP in constant 1985 dollars (kgoe/US$1,000 1985). SOURCE: Grübler (1991).

4) reveals the large differences in the policies and structures of energy systems among countries. For example, though Japan and France have both achieved high degrees of decarbonization, they have followed disparate routes. At the global level, the long-term overall reduction in carbon intensity per unit of value from both factors totals about 1.3 percent per year since the mid-1800s.

The major determinants of energy-related carbon emissions can be represented as multiplicative factors in a simple equation. Placing carbon emissions on one side, on the other we have population growth, per capita value added, energy consumption per unit of value added, and carbon emissions per unit of energy consumed (Yamaji et al., 1991). As we have seen, the last two terms in this equation are decreasing globally. However, their decline is counteracted by rising values for the preceding terms, population and economic activity, resulting in an overall global increase in energy consumption and carbon emissions.

The world's global population is currently increasing at a rate of about 1.6 percent per year. The longer-term population growth rate since 1800 has been about 1 percent per year. Most population experts predict at least another doubling during the next century (see United Nations, 1992, and Vu, 1985; see also Kates, this volume). Economic activity has been increasing in excess of global population growth since the beginning of industrialization, made possible by the productivity increases referred to at the outset of this essay. In recent decades global economic growth, stirred by both population and productivity gains, has proceeded at about 3 percent per year. Subtracting 1.3 percent for decarbonization, the result is that global carbon emissions have been increasing at about 1.7 percent per year. A continuation would imply a doubling of emissions in about forty years. Fearing such an increase, we must examine in detail the differing paths to decarbonization to see what the limits of the process might be.

DECONSTRUCTING DECARBONIZATION

An examination of five countries—China, France, India, Japan, and the United States—furthers our understanding of the decarbonization process (Nakićenović, 1996). These countries represent diverse economic and energy systems and life-styles as well as a significant share of the world's energy use. The United States has one of the highest energy intensities of all the industrialized countries, and the highest per capita energy consumption in the world. France and Japan have among the lowest energy intensities in the world, but for different reasons, as we shall discuss. China and India are rapidly developing and still replacing traditional energy sources with commercial ones, and thus they exhibit very high energy and carbon intensities. Together, the five countries account for about 45 percent of global primary energy consumption and more than 40 percent of energy-related carbon emissions.

To determine more precisely the various causes and determinants of the decreasing carbon intensity of energy, we disaggregate the energy system into its three major constituents: primary energy consumption, energy conversion, and final energy consumption. Primary energy consumption embraces the requirement for original resources such as coal, crude oil, and uranium. Final energy refers to the gasoline pumped into a car's fuel tank, the electricity for powering a room air conditioner, or firewood if used directly for cooking or heating. Primary energy, such as coal, is rarely consumed in its original form in a household or office but rather is converted into electricity, fuel, and heat. Thus, final energy, which is consumed directly, in some sense represents best the actual energy requirements of the economy and individual consumers.

In fact, neither primary energy consumption nor conversion is transparent to consumers. For example, the production process for electricity is invisible to most consumers. Because electricity itself is carbon-free, it does not emit carbon (or soot, sulfur dioxide, and other pollutants) at the point of consumption. However, carbon can be emitted in converting primary energy forms into electricity. To a lesser degree this is also true of other forms of final energy, such as oil products. Although the carbon emissions per liter of diesel or gasoline finally used in a truck are basically the same throughout the world, the carbon emissions produced in converting different grades of crude oil into the two products can vary substantially.

To deconstruct the constituent decarbonization rates of the energy system, we make three assumptions. First, the carbon intensity of primary energy is defined as the ratio of the total carbon content of primary fuels to total primary energy consumption for a given country. Second, the carbon intensity of final energy is defined as the carbon content of all forms of final energy divided by the total final energy consumption. The third assumption is that the carbon intensity of energy conversion is the difference between the two intensities just described. So, for example, the carbon intensity of primary energy runs high when wood and coal supply most of the fuel. The carbon intensity of conversion runs high when coal burns to make most of the electricity and when the conversion (or transmission and distribution) system itself is wasteful. Efficiency improvements in the energy system mean that less primary energy is consumed per unit of final energy; lower conversion losses therefore result in lower carbon emissions. The carbon intensity of consumption runs high when the final consumer cooks with coal or travels by gasoline and when end-use devices are inefficient.

Let us now compare the carbon intensities of final, primary, and conversion energy for the United States, Japan, France, China, and India in recent decades (Figures 5 through 7). Steady reductions in the carbon intensity of final energy in all five countries stand out above all. On average, the three industrialized countries have spared about 20 percent since 1960, while the pair of developing countries have cut back about 15 percent since the early 1970s. The reductions converge tightly in the three industrialized countries. The gap between the developed and the developing countries is also slowly narrowing because of the slightly more rapid declines in intensity in the latter.

The major reason for the decarbonization of final energy is the increasing share of electricity in final energy throughout the world. The percentage of global primary energy used to create electricity has climbed during this century from 5 in the year 1910 to 20 in 1950 to about 35 in 1990. A second reason is that the average mix of other fuels consumed for final energy has a decreasing carbon content, that is, greater shares of oil products and natural gas. Accordingly, these products also have a higher hydrogen content, a point that will be discussed in the final section of this essay.

The carbon intensity of primary energy has also fallen in all five countries, though only very slightly in the United States, where coal has retained its strong role. The carbon intensities of conversion give a completely different picture, however. The diversity in the development and structure of the energy systems of the five countries becomes apparent. In the developing countries, the carbon intensity of conversion has increased, while in France it dropped sharply; in the United States and Japan the conversion intensity initially rose before declining during the latter part of the period analyzed.

Should China and India continue to rely heavily on coal as their primary

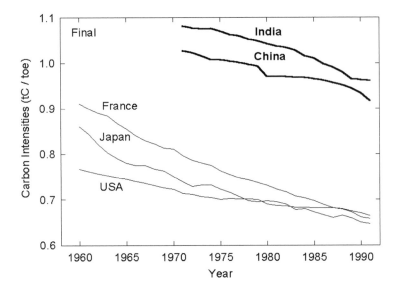

FIGURE 5 Carbon intensities of final energy, expressed in tons of carbon per ton of oil equivalent energy (tC/toe).

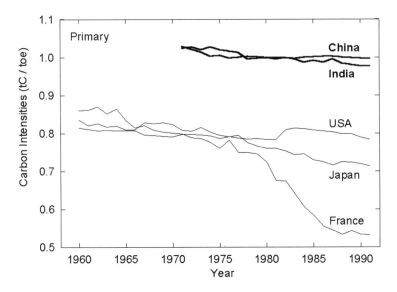

FIGURE 6 Carbon intensities of primary energy, expressed in tons of carbon per ton of oil equivalent energy (tC/toe).

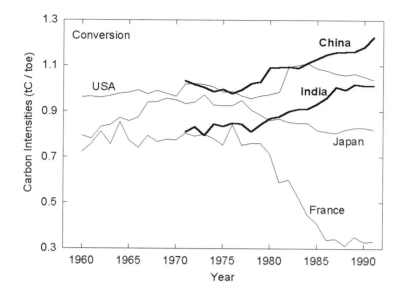

FIGURE 7 Carbon intensities of energy conversion, expressed in tons of carbon per ton of oil equivalent energy (tC/toe).

source of energy, continuing to lessen the carbon intensity of primary energy in these countries will prove difficult. In fact, sometime in the next century the downward trend in the carbon intensity of primary energy could reverse itself, caused by an even higher share of electricity in end use but generated with coal. Alternatively, China and India could restructure their energy systems to make increasing use of natural gas or nuclear energy and other zero-carbon options. Such shifts would align their energy systems with those of the more industrialized countries.

Focusing on the United States and Japan, we see that the carbon intensity of primary energy exceeds that of final energy, with conversion intensity the highest of the three. While final carbon intensity decreases somewhat faster in Japan (about 0.8 percent per year) than the United States (about 0.5 percent per year), the difference in the conversion intensities is much more dramatic. In both countries the changes in the carbon intensity of energy conversion are erratic, especially compared to the steady improvements in final intensities. The overall reduction of carbon intensity in Japan stems primarily from improvements in energy efficiency and, to a lesser degree, from the replacement of carbon-intensive energy forms.

France provides a contrast. Here, the rapid introduction of nuclear energy since the mid-1970s has led to higher rates of decarbonization of primary energy and of conversion (because an increasing share of electricity is produced without carbon emissions) than of final energy. This strategy to achieve low carbon emissions is completely internal to the energy system and fundamentally decoupled from the consumer. Nevertheless, the relatively smooth improvement in final carbon intensity is similar to that observed in Japan and the United States.

China and India present a different picture, though they resemble one another. The three energy ratios and their evolution are similar in these countries despite their many social and cultural differences, as well as those differences that may be attributed to the varying development paths of planned and market economies. In both countries, the carbon intensity of primary energy is diminishing slightly. The carbon intensity of final energy, on the other hand, decreases at rates comparable to those observed in industrialized countries. In India, the faster decarbonization of final energy is due to the replacement of traditional fuels by commercial energy forms. For example, the use of biomass (mainly wood that is not replaced by a new forest) is more carbon intensive than using either kerosene or bottled gas. The difference in carbon intensity between electric lighting (especially if efficient light bulbs are used) and traditional illumination is even more pronounced. In any case, the developing economies are undergoing basically the same process of decarbonizing final energy use as the most developed countries.

In the industrialized countries, the decarbonization of final energy consumption has been accompanied by additional structural changes in the energy system. These led to improvements in decarbonization in the energy system itself, as demonstrated by the downward trends in the carbon intensity of conversion. In contrast, China and India have not undergone this transition. Their energy systems depend heavily on coal, whereas most industrialized countries have in large measure replaced coal with less carbon-intensive sources, even in electricity production. As a consequence of their dependence on coal, both China and India show rapid increases in the carbon intensity of conversion. Should a transition to a lower carbon intensity in developing countries not occur in the coming decades, the likely reductions in carbon emissions in the industrialized countries will be offset, hampering efforts to halt the global increase in carbon emissions.

In sum, determining decarbonization only as the ratio of total carbon emissions per unit of primary energy consumption may veil the interaction between the energy system and the economy. As the structure of an energy system changes, so does the carbon intensity of its three constituent parts. The actual forms of final energy demanded and consumed matter greatly in the logic of decarbonization. Because electricity and heat contain no carbon, the carbon intensity of final energy is generally lower than the carbon intensity of primary energy. In addition, its rate of decrease exceeds that of primary energy because of the increasing share of electricity and other fuels with lower carbon content, such as natural gas, in the final energy mix. At the level of final energy, decarbonization is a durable, pervasive phenomenon. The likely explanation is a congruence in consumer behavior and preferences as expressed in the structure of final energy over a wide range of income and developmental levels.

THE ELEMENTAL EVOLUTION

We have seen the increasing needs for electricity and hydrogen-rich forms of *final* energy. Can these be reconciled with the relatively slow and often opposing changes in the structure of energy systems and the *primary* energy supply? The historical replacement of coal by oil, and later by natural gas, at the global level shows the way. The well-documented evolutionary substitution of sources of primary energy suggests that natural gas and later carbon-free energy forms will become the leading sources of primary energy globally during the next century (Ausubel et al., 1988; Grübler and Nakićenović, 1988; Marchetti and Nakićenović, 1979; Nakićenović, 1990).

The competitive struggle between the five main sources of primary energy—wood, coal, oil, gas, and nuclear—has proven to be a dynamic and regular process that can be described by relatively simple rules. A glance reveals the dominance of coal as the major energy source between the 1880s and the 1960s after a long period during which fuelwood and other traditional energy sources led (Figure 8). The mature coal economy meshed with the massive expansion of railroads and steamship lines, the growth of steelmaking, and the electrification of factories. During the 1960s, oil assumed a dominant role in conjunction with the development of automotive transport, the petrochemical industry, and markets for home heating oil.

The model of energy substitution projects natural gas (methane) to be the dominant source of energy during the first decades of the next century, although oil should maintain the second largest share until the 2020s. Such an exploratory look requires additional assumptions to describe the later competition of potential new energy sources such as nuclear, solar, and other renewables that have not yet captured sufficient market shares to allow reliable estimation of their penetration rates. In Figure 8 it is assumed that nuclear energy will diffuse at rates comparable to those at which oil and natural gas diffused half a century earlier. Such a scenario would require a new generation of nuclear installations; today such prospects are at best questionable. This leaves natural gas with the largest share of primary energy for at least the next fifty years. In the past, new sources of energy have emerged from time to time, coinciding with the saturation and subsequent decline of the dominant competitor. In Figure 8, "Solfus" represents a major carbon-free energy technology, such as solar or fusion, that could emerge during the 2020s at the time when natural gas is expected to reach the limits of its market niche.

The unfolding of primary energy substitution implies a gradual continuation of energy decarbonization globally. Figure 9 shows how the ratio of hydrogen to carbon atoms in the world fuel mix has changed as a result of primary energy

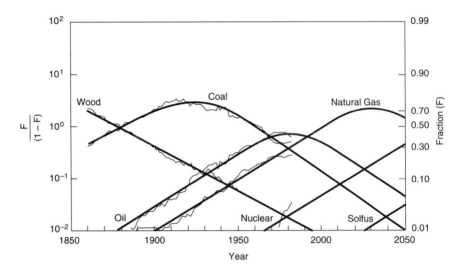

FIGURE 8   Global primary energy substitution from 1860 to 1982 and projections for the future, expressed in fractional market shares (F).   NOTE: Smooth lines represent model calculations and jagged lines are historical data. "Solfus" is a term employed to describe a major new energy technology, for example, solar or fusion. SOURCES: Grübler and Nakićenović (1988) and Nakićenović (1990).

substitution. If natural gas becomes the dominant source of energy, this ratio can be expected to approach the level of four hydrogen atoms to one carbon. Improvements beyond this level would have to be achieved by the introduction of noncarbon energy sources and by the sustainable use of biomass.

A methane economy offers a bridge to the noncarbon energy future consistent with both the dynamics of primary energy substitution and the steadily decreasing carbon intensity of final energy. As nonfossil energy sources are introduced into the primary energy mix, new energy conversion systems would be required to provide zero-carbon carriers of energy in addition to electricity. The ideal candidate is pure hydrogen, used as a gas or liquid. Hydrogen and electricity could carry virtually pollution-free and environmentally benign energy to end users in a carbon-free energy system.

To the extent that both hydrogen and electricity might be produced from methane, the carbon separated as a by-product could be contained and stored, probably in underground caverns. As the methane contribution to the global energy supply reaches its limit and subsequently declines, carbon-free sources of energy would take over, eliminating the need for carbon handling and storage. This would conclude the global trend toward decarbonization and the resulting major transformation of the industrial ecosystem. The emergent system could accommodate cleanly the foreseeable levels of population and economic activity.

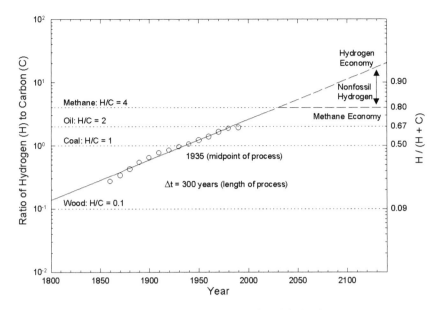

FIGURE 9   Ratio of hydrogen (H) to carbon (C) for global primary energy consumption since 1860 and projections for the future, expressed as a ratio of hydrogen to carbon (H/C). SOURCE: Ausubel (1996) and Marchetti (1985).

In fact, an energy system of the distant future that relies on electricity and hydrogen as the complementary energy carriers would also advance dematerialization. Hydrogen has the lowest mass of all atoms, and its use would radically reduce the total mass flow associated with energy activities and the resulting emissions. Electricity is free of material emissions, and the only product of appropriate hydrogen combustion is water. Thus, decarbonization not only contributes to dematerialization but is also consistent with the emergence of new technologies that hold the promise of high flexibility, productivity, and environmental compatibility. Weighty carbon is a poor match for the evolving final energy demands of modern societies. Fortunately, decarbonization has asserted itself already as a widespread, long-term development driven by deepening, strengthening forces.

REFERENCES

Argote, L., and D. Epple. 1990. Learning curves in manufacturing. Science 247:920–924.
Ausubel, J. H. 1996. Can technology spare the Earth? American Scientist 84(2):166–178.
Ausubel, J. H., and A. Grübler. 1995. Working less and living longer: Long-term trends in working time and time budgets. Technological Forecasting and Social Change 50:113–131.

Ausubel, J. H., A. Grübler, and N. Nakićenović. 1988. Carbon dioxide emissions in a methane economy. Climatic Change 12:245–263. Reprinted as International Institute for Applied Systems Analysis RR-88-7.

Christianson, L. 1995. Diffusion and Learning Curves of Renewable Energy Technologies. WP-95-126. Laxenburg, Austria: International Institute for Applied Systems Analysis.

Grübler, A. 1991. Energy in the 21st century: From resource to environmental and life constraints. Entropie 164/165:29–33.

Grübler, A., and N. Nakićenović. 1988. The dynamic evolution of methane technologies. In The Methane Age, T. H. Lee, H. R. Linden, D. A. Dreyfus, and T. Vasko, eds. Dordrecht, Netherlands, and Laxenburg, Austria: Kluwer Academic Publishers and International Institute for Applied Systems Analysis.

Maddison, A. 1991. Dynamic Forces in Capitalist Development: A Long-Run Comparative View. New York: Oxford University Press.

Marchetti, C. 1985. Nuclear plants and nuclear niches. Nuclear Science and Engineering 90:521–526.

Marchetti, C., and N. Nakićenović. 1979. The Dynamics of Energy Systems and the Logistic Substitution Model. RR-79-13. Laxenburg, Austria: International Institute for Applied Systems Analysis.

Nakićenović, N. 1990. Dynamics of change and long waves. In Life Cycles and Long Waves, T. Vasko, R. Ayres, and L. Fontvieille, eds. Berlin: Springer-Verlag.

Nakićenović, N. 1996. Decarbonization: Doing more with less. Technological Forecasting and Social Change 51:1–17.

United Nations. 1992. Long-Range World Population Projections: Two Centuries of Population Growth 1950–2150. New York: United Nations.

Vu, M. T. 1985. World Population Projections. Baltimore: Johns Hopkins University Press.

Williams, R. H., E. D. Larson, and M. H. Ross. 1987. Materials, affluence, and industrial energy use. Annual Review of Energy 12:99–144.

Yamaji, K, R. Matsuhashi, Y. Nagata, and Y. Kaya. 1991. An Integrated System for CO_2/Energy/GNP Analysis: Case Studies on Economic Measures for CO_2 Reduction in Japan. Paper presented at the Workshop on CO_2 Reduction and Removal: Measures for the Next Century, International Institute for Applied Systems Analysis, Laxenburg, Austria, March 19–21, 1991.

Technological Trajectories and the Human Environment. 1997.
Pp. 89–109. Washington, DC: National Academy Press.

Life-styles and the Environment:
The Case of Energy

LEE SCHIPPER

When we speak of pressures on the natural environment, we should speak more about home loans, old-age income, and women drivers, more about shrinking households and all-night shopping, and perhaps less about coal mines and pulp mills. In this essay I will argue that the precise nature of the demands for services that we collectively create increasingly shape environmental change. While the ways we farm, mine, and manufacture surely transform the environment, the end points of economic activity, what we consume, how we actually live—"life-styles," in short—are no less important.

Because significant energy use accompanies or allows for most human activities, I choose energy use to indicate the overall level of environmental disruption. Of course, environmental impacts arise also from the volume and selection of materials we employ, the ways we alter water courses and habitat, and numerous other activities. Moreover, I do not want to attach an environmental stigma to energy, which we use precisely because it serves us so well. But, for any given mix of fuels and remediative techniques, the level of energy use does provide a reasonable proxy for the environmental impact of human activities. We also know that changes in technology, both to improve efficiency and to shift energy sources, can reduce emissions and other unwanted fallout per unit of service (see Nakićenović, this volume). So can changes in the levels of services demanded, stimulated by changes in incomes, prices, or, as I shall emphasize, life-styles.

To give human meaning to the link between energy and life-styles, consider the atmospheric emissions per capita of carbon dioxide, the gas that is most threatening climatic change, in 1973 and in 1990 from the energy-using activities of the households of five countries (Figure 1) (Sheinbaum and Schipper, 1993).

89

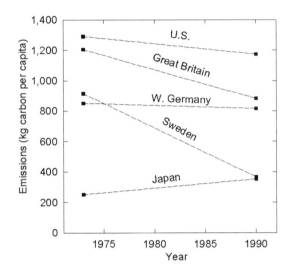

FIGURE 1 The changing household emissions of carbon in five countries. NOTE: These emissions convert to the greenhouse gas, carbon dioxide.

The figure demonstrates the diversity of social circumstances and the means to increase energy services while environmental fallout moderates, as well as our ability to reveal connections between emissions and specific activities.

FROM PRODUCTION TO PLEASURE

What a society makes still matters. In fact, changes in the mix of productive activities forming the economy have profoundly altered energy use in recent decades. Between about 1970 and 1990, structural changes reduced energy consumption by 10 percent or more in the United States, Japan, and West Germany, compared with demand that would have arisen had the mix of production remained constant (Howarth et al., 1993; Schipper et al., 1992a). The main change has been a reduction in the role of manufacturing, which has accounted for 30 to 40 percent of energy use in industrialized countries. Improvements in the energy efficiency of manufacturing have also significantly tempered the growth in energy demand (Schipper, 1993). Energy-use efficiency is still improving by 1 to 2 percent per year in manufacturing.

Energy-use efficiency has improved more slowly in households, services, and transport. In fact, changes in the ways consumers use energy—for comfort at home and personal mobility, for example—have raised energy use in all these sectors. The sum of the recent changes in energy demand is essentially a shift in

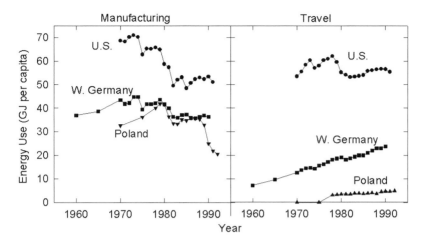

FIGURE 2 Contrasting trends in energy use for manufacturing and travel.

prominence from producers to individual consumers and to the collective consumption of the service industries (retail stores, office buildings, and the like).

This shift from "production to pleasure" has been distinct in Germany and elsewhere, and in formerly planned economies, such as Poland, it has been drastic (Figure 2) (Schipper et al., 1989). The rapid changes after the fall of the old communist regime in 1988 and 1989 shrank heavy industry relative to other activities in the Polish economy, while travel climbed, beginning from a very low base. Energy use and restructuring in all the former Soviet economies has essentially resembled that of Poland (Cooper and Schipper, 1992). For a time, the shift was barely discernible in the United States because energy savings from smaller private cars stabilized growth in energy travel demand; however, recent trends toward larger personal vehicles and higher speed limits are now causing growth to resume.

Excepting infrequent, dramatic events of the kind that occurred in Central and Eastern Europe, energy demand does not change much in the short term unless energy prices or incomes change, in turn causing changes in the demand for energy services. In the medium term, however, the systems converting energy to services, such as hot food or cool air, may be modified, renovated, or replaced, and these changes in capital stock and associated techniques allow for enormous changes in energy requirements per unit of service. Individuals also do not generally change their energy-demanding behavior rapidly, but they can and do modify their comfort and mobility patterns somewhat. The changes may be temporary, as was the case in the United States in response to the oil price shocks of the 1970s. In the long term, because technologies, human behavior, and populations all change, quite different patterns and levels of energy use can emerge. The potential for technological changes that save energy and other resources has received much attention (Schipper, 1993; Schipper et al., 1992b). In this essay, the focus is on life-style and the related factors that affect the level of energy services demanded—raising or lowering resource consumption, or changing the "resource intensity" with which incomes are disposed.

Life-style is a rubric that covers numerous activities in which individuals engage. These include spending for personal consumption, owning and using goods, and travel. Life-styles reflect social and demographic characteristics, including the age and employment status of individuals and the size of families. Life-style choices may occur at the level of the individual, the household, or much larger groups; they relate primarily to money and time, to how much time, for example, is spent outside the home. Life-style characteristics are not independent; families with small children, for example, may need to spend more time at home than those without children. Life-styles evolve in response to changes in income and taste.

CONSUMPTION EXPENDITURES

Savings as well as present and projected income constrain the goods and services that households are able to purchase in any given year. How households allocate their incomes among goods, services, and savings reveals economic preferences, and consumer expenditure data show how much money average households spend for various goods and services. Since 1970, households in many countries have spent substantially more for transport, primarily on automobiles and other travel services (Schipper et al., 1989), and this change favors increased energy consumption.

Difficulties in translating money into quantities of activity limit the direct use of expenditures to assess impacts on energy and the environment. For example, what kinds of equipment were consumers purchasing when they spent for "household appliances"? How much energy did such appliances require? If consumers were spending more money for meals taken out of the home, how many kilometers were they traveling to secure those meals? Was an additional expenditure for recreational goods used in the home, at a neighborhood gymnasium, or in more distant ski country? Consumer expenditures present an aggregate picture that has to be connected to the infrastructure of consumption, reflecting the ownership of and access to energy-using goods.

The characteristics of the stock of energy-using goods in personal and social infrastructures indicate how consumers use energy. During the 1950s and 1960s in the United States and during the 1960s and 1970s in Europe, ownership of central heating systems increased steeply (Figure 3), while home area increased as well (Figure 4). Automobile ownership grew in Europe, moving toward US levels (Figure 5). Because central heating systems tend to use twice as much energy per square meter of floor space as room heaters, and because personal transport uses more energy per passenger-kilometer than mass transit, personal energy use rose rapidly in America and Europe. The increase in home heating and car ownership alone probably doubled per capita household energy use in Western Europe between 1960 and 1990. The increases in ownership raised energy demand more than the changes in the use of the systems, such as the hours of heating in homes equipped with central heat. By the late 1980s, growth in ownership of many kinds of equipment slowed both in the United States and in other high-income countries. However, the acquisition of key energy-intensive personal goods continues in Japan up to the present.

Greater income has not, however, always led to proportional increases in energy use, principally because the goods subsequently purchased did not necessarily consume energy at the high levels of heating or cooling systems. Moreover, as the ownership of equipment approaches saturation—with every household owning a given device, and every person with a driver's license having at least one car—the characteristics of these products and their overall utilization become increasingly important in determining energy use. Because successive vintages of the same goods tend to improve in efficiency, increases in the level of ownership have had a declining impact on the growth of energy use through the 1980s and 1990s. Of course, new energy-intensive appliances may appear on the market. As yet, the energy intensity of the sharply rising number of information-handling devices such as computers and printers remains unclear.

Data for newly industrializing countries, former socialist economies, and developing countries show a large gap in their ownership of goods as compared with the United States and Western Europe. The gap between Eastern and Western Europe was narrowing even before the political and economic changes of the late 1980s. Fed with costly imports of used Western cars, the gap in car owner-

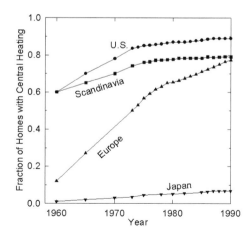

FIGURE 3   The rise of central heating.

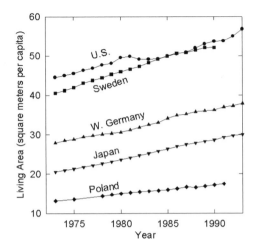

FIGURE 4   The expansion of residential space.

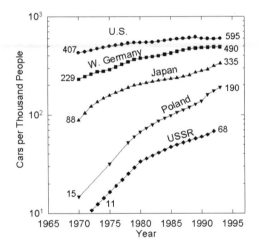

FIGURE 5   The growth of car populations.

ship, which accelerated after 1989 in Poland, continued to shrink even as the economies of the former East bloc were in collapse—suggesting strongly that individuals will do almost anything to acquire the personal mobility previously rationed or restricted (Meyers et al., 1993) Recent construction in Poland and other countries points to an analogous reach for housing space.

TIME AND DISTANCE

Surveys reveal how much time individuals spend in various activities during the course of the typical day or week (Gershuny and Jones, 1987; Szalai, 1972). Time budgets classify activities by purpose, such as paid work, leisure, and travel, and sometimes by location, at home or at work. A group of Americans surveyed in 1985, for example, spent about 6 percent of their weekly time traveling, 14 percent at work, and 70 percent at home, half in sleep (Robinson et al., 1988). Changes in the time spent performing any given activity require trade-offs with other pursuits.

People appear to be gradually changing the ways they spend their time. For example, leisure time in the United States was slightly lower in the 1980s than in the 1970s, but slightly higher in Europe. Surveys suggest Americans, Swedes, Norwegians, and Danes are all spending more time away from home (Carlsson, 1989; Ketterød, 1992; Mogensen, 1990). To see whether this trend is reflected in mobility, we next review measures of distance traveled.

Countries differ much more significantly in per capita travel when measured in distance rather than in time, indicating a variation in average travel speed. Speed, of course, demands energy. Americans continue to travel almost twice the passenger-kilometers of Europeans (Figure 6). In the United States in the 1960s, the level of travel already exceeded the level of much of Europe today, yet US travel continues to increase. In North America, the automobile accounts for 85 percent of personal miles traveled, and in Europe more than 80 percent. Air traffic continues to increase, providing most of the remaining distance traveled in the United States. Distances traveled by bus and rail have remained at about 3 percent of total US travel since the early 1970s, but their share is falling in Europe.

At present, Americans and Europeans spend about 1 minute traveling for every 4 to 5 minutes spent in out-of-the-home work, leisure, and shopping activities (Gershuny and Jones, 1987). Americans go faster for a slightly longer time; consequently they travel considerably farther. Russians travel far shorter distances than do Americans or Europeans, but they face commute times as long as Americans because most must travel circuitous routes on mass transit, even where direct distances to work are small.

The high level of US travel and its continued growth make improbable the view that per capita travel distance has reached saturation anywhere. In fact, no country shows evidence that travel distance has saturated.[1] Suggesting the contrary, people have been spending somewhat more time away from home. Unless people travel shorter distances to services, work, and leisure, or make fewer trips, energy demand for transport will increase in the absence of a change to a markedly more energy-efficient travel technology.

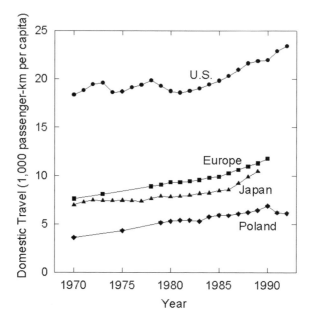

FIGURE 6　Average yearly distance traveled by all modes of transport in domestic traffic.

POPULATION AND HOUSEHOLD CHARACTERISTICS

Demographic characteristics affect the use of homes, commercial buildings, and transport services. The household is the most important unit to consider, because most energy-using goods in homes or on the road are shared by several household members. Changes in household size and age distribution, the nature of employment, and urbanization affect the use of energy-consuming goods.

Energy use in households increases more slowly than household size. In fact, it can reasonably be represented by the square root of this quantity. Smaller households use more energy per capita than larger ones when comparisons are made to control for income, fuel type, and other like factors. This relationship matters greatly given the worldwide trend toward smaller households, most conspicuous in developed countries. Couples now marry later and have fewer children. Children tend to leave the family home earlier in life than they did in the past, increasing the number of single-person households. And, people are living longer. Between 1960 and 1985 household size fell from 2.9 to 2.4 in Sweden, from 3.6 to 2.6 in the Netherlands, from 3.4 to 2.8 in the United States, and from 4.5 to 3.2 in Japan. Roughly speaking, the declines can be expressed as increases in the number of housing units per capita; these increases caused per capita household energy use to rise by 15 to 25 percent. Considering larger household

sizes in the countries of the former Soviet Union (3.5) and in developing countries (well over 4 in the newly industrializing countries and even higher in less-
developed countries), shrinking households loom as a profound force making for
higher energy use.

The ages of household members also determine energy use. For example, in
the United States "elderly singles" (sixty years or older), with only 10 percent at
work, will tend to stay at home a good deal more than younger singles. Not
surprisingly, these single elderly people on average use more energy than younger
singles for both heating and appliances, though the low-income elderly use less
(Diamond, 1987). Because elderly singles own fewer cars and drive considerably
less than others, there is some savings here.

Generally, energy use in the home evolves as the family moves through its
life cycle (Schipper et al., 1989). Residential energy use typically increases with
the birth of children and then rises slightly, peaking when the children are in their
teens (Gladhart et al., 1986). Comparisons of family types show that "couples
with young children" use less energy than "couples with old children," even
when family size is virtually the same. Because income tends to increase as
family members move through their career paths, family-cycle and income effects may combine, leading the growing family to move into a larger home, and
thereby increasing space-heating energy use. Because driving rises with age and
then falls for both men and women (Figure 7), the energy-use cycle is reinforced,
though overall societal mobility is rising over time.

Great differences in per capita energy use reflect differences in household
composition. In the United States, the typical family with children (married parents, 2.1 children) uses the least energy per capita, but comprises a declining
share of the total number of households. Single-person households, households
consisting of unrelated persons, single-parent households, and elderly households
are increasing almost everywhere. This shift tends to raise per capita energy use
in households and transport as well.

Subtle changes in energy use can occur as the demographic structure of
society evolves. Fewer children means more time available for leisure and also
for work for women. Indeed, female participation in the labor force has increased
steadily since 1950 in the United States and in other developed countries. For
families where both parents work, automobiles allow parents to visit schools, run
errands, and commute to workplaces within very tight time constraints. Not surprisingly, two-worker households, increasing today in all countries, have significantly different driving patterns than one-worker households.

The growth in the number of elderly as a share of the total population in
industrialized countries also matters for energy. In 1985 the US elderly spent one
hour more per week traveling for leisure than the national average. Although they
drive far less than younger cohorts, the distances traveled are increasing over
time. In addition, people over the age of sixty-five spent five more hours than the
national average per week using electronic media in the home. Extrapolation of

FIGURE 7   Miles driven yearly by Americans, by age group and gender.

present patterns would suggest that energy use for driving will diminish as the population ages, while home energy use will grow as more one- and two-person "elderly" households are formed.

The life-styles of the "elderly" are themselves changing. A new generation of active retirees is forming, whose energy-related behavior is not known. Tomorrow's energetic retirees may carry with them the mobility patterns of their younger years, continuing to live in homes originally built to house families with two or three children or in sunbelt homes that require substantial air conditioning. Although these retirees will eventually swell the ranks of the less-energetic geriatrics (those in their high eighties and nineties), their energy use may remain high on a per capita basis, particularly if they keep their previous energy-using capital (homes and cars) and habits (heating, cooling, driving, flying).

The demographic changes intertwine not only with income effects but also elements of public policy. Indeed, public policy influences income indirectly by controlling the cost of services through subsidies or taxes. Public child-care facilities and liberal maternity-leave policies, for example, mean that more women are able to work. The continued growth of the elderly segment of society will depend on our ability to pay for high standards of medical care, and the increased number of people living alone must reflect their being able to afford such independence. The ability of the elderly to continue to live in their family homes depends on how well their savings and pensions, including social security, support them in retirement. One reason for larger household size in the former Soviet Union, for example, is simply that parents live with children and grandchildren out of financial need, which in turn depends on social norms but also on government policies.

CUSTOMS AND LAWS

Culture, custom and tradition, and government policies all affect energy use by influencing how and where people live, how they move about, what they do. The role of women in society, to take an area of major change, illustrates the powerful relationships between social mores and energy use. More women currently participate in activities previously dominated by men. A greater number are employed and are obtaining higher education, delaying and even forgoing marriage and childbearing. More single people and more childless couples mean more households and thus greater energy use.

A less noticed change with respect to women relates to their driving habits. At any given age, except above age sixty-five, the share of women with drivers' licenses almost equals the share of men. Increases in women driving, particularly for work-related purposes, was an important reason for the increased driving per capita in the United States between 1969 and 1990, as suggested in Figure 7 (Federal Highway Administration, 1982, 1986, 1991). If women in the United States still drive only half as much as men, the gap is narrowing (Hu and Young, 1993). Closing the gap both in participation and distances driven could increase gasoline use significantly in many countries.

Where people travel is a function of social traditions that often limit the times when specific facilities are open. Longer opening hours for shops, places of work, and places of entertainment has led to more off-peak use of transport and greater utilization of buildings and, therefore, heating, lighting, and air conditioning during these extra hours. When these services opened only during restricted hours—still the case for shops in much of Europe—demand for transport peaked accordingly, with little demand during off-peak hours. In the United States, "seven-eleven" shopping hours have become common in many large downtown stores and in shopping malls, as well as in convenience stores. Attitudes and policies toward restaurants and entertainment also affect the likelihood that people will spend leisure time out of their homes. Company and tax policies favoring entertainment expenses influence how much people choose to go out.

Extended hours allow a given building to be exploited for longer periods. Whether this extra use raises or lowers energy consumption per unit of income generated (or per visitor or employee) depends on the ratio of the extra energy required for keeping the building open to the additional business generated. But compared to erecting two separate buildings to accommodate a given level of business, increasing the opening times of a single building for the same business would appear to reduce the fuel and electricity required to support a unit of business activity.

Extending the number of hours also entails extending working hours, leading to longer workdays or more part-time jobs. Changes in working hours influence energy use both in transport and in buildings. Shorter or more flexible working hours may encourage combined trips for shopping and work. At the same time, however, increased part-time work means a greater number of commuters. One social change that may reduce occupancy in service-sector buildings is the increased use of the home for conducting paid work (telecommuting) and accessing services (through, for example, shopping-by-mail catalogs, or electronically by television or computer). This change alters both commuting and home energy-use patterns, reducing occupancy in places of employment and services.

Not all "work" appears in national accounting systems. "Informal" or unpaid work for oneself (do-it-yourself) or for others (baby-sitting and bartering) may be increasing today (Bonke, 1986; Sanne, 1988). These changes in work patterns may change the utilization of buildings, if people choose to stay home more (Sanne, 1988). The rise in informal production may also affect travel: when services or do-it-yourself activities are important, people tend to remain in their neighborhoods, where they may know their neighbors and obtain trusted services. Similarly, "producing" services with inexpensive equipment, rather than paying for these (renting a film for the video-cassette recorder instead of going to the movies), also represents a significant shift in the way goods and services are produced (Gershuny and Miles, 1987).

In short, the very notion of how and where individuals "produce" in society may be changing, and with this may come a change in the amount, type, and location of energy use. The principal impact appears to be the transference of work from manufacturing or services into the home. The energy implications so far are small, but they could become significant if the energy demands of the information economy become large and pervasive.

Taxation policies may also affect personal energy use. For example, tax benefits for payment of mortgage interest stimulates the construction or purchase of single-family dwellings. In the United States, tax rules permit unlimited deduction of mortgage interest payments from taxable income; this same benefit is limited in most European countries. While US authorities permit almost no deductions for commuting costs, these are directly deductible in Sweden, or indirectly subsidized through light tax treatment of company-provided cars or company-subsidized transit tickets in the United Kingdom and Germany. Changes in such rules demonstrably and rapidly alter the type and location of homes built, the kinds of cars purchased, and the ways in which they are used. The result is larger homes built farther apart and, in countries with liberal company-car policies, larger cars and additional driving. During most of the 1970s and 1980s, about one-third of all new cars bought in Sweden were provided for employees by companies for personal use (Schipper et al., 1993); on the average, they were considerably heavier and more powerful than the so-called private cars.

A US-Sweden comparison offers other insights into how policies and customs may have an important indirect effect on energy use. While tax policy favors home ownership in both countries, housing policy in Sweden for many decades favored the construction of apartments (Schipper et al., 1985), while no such policy has taken hold in the United States. As a result, Swedish cities are more compact, and the proportion of apartments in the total dwelling stock, many owned by their occupants, is higher in Sweden than in the United States or in most other European countries. Still, the space heating consumption in apartments in Sweden is higher, on a per capita or unit area basis, than in single-family dwellings (Schipper, 1984), and this is because few apartments are metered individually for actual heat consumed. Thus, there is no clear evidence that housing policies in Sweden have led to energy savings in households as a result of the differences in the kinds of dwellings built. Some would argue, however, that the high share of apartments in compact settlement patterns encouraged the use of mass transit, which held steady in Sweden during the 1970s and 1980s. The fact that Sweden has the highest ownership of second or "summer" (now often winterized) homes is probably related to the high share of apartments in the ordinary housing stock. As Fredbaeck (1979) noted, fuel and electricity used in these homes is small compared to what is used in the principal homes, but the fuel used commuting back and forth turns out to be significant. As compensation for living in compact settlements in cities, Swedes travel back and forth frequently to their summer homes.

National policies affecting the social security system and care for the elderly may have large energy implications. In Scandinavia, for example, liberal pensions permit retirees to travel and lead active independent lives; concerns about retirement force Japanese families to save more and live together longer. Changes in the leisure time that societies formalize through paid vacation and holidays also affect energy demand. If people choose to work fewer hours, they may find themselves with significantly more free time. Scandinavians, for example, have as much as six weeks paid vacation; their paid holidays are substantially greater than any enjoyed by workers in the United States. A key issue with respect to energy is whether these vacations involve travel, staying at home, or making a single trip to a summer home. Free time in general has become much more travel-dependent. It is obvious that the evolution of customs and policies has changed how and where money is both earned and spent. On balance, governmental policies have encouraged personal energy use.

If higher incomes, demographic changes, and government policies feed the demand for energy, what has been the net effect, including changes in efficiency, on the demand for energy in major sectors related to changes in gross domestic product in recent decades? Our analyses of numerous countries suggest that quite different conditions prevail in manufacturing, travel, and household energy services (Howarth et al., 1993; Schipper et al., 1992b; or Schipper et al., 1993). The calculations for manufacturing show clear, sizable declines; Japan needs only about two-thirds the primary energy services in manufacturing that it used per dollar of GDP in the 1970s (Figure 8). In travel, only the United States has maintained a lower ratio since the 1973 oil price shock (Figure 9). For household energy services, the trend was steeply upward through the 1970s for many countries including Germany (Figure 10) (Sheinbaum and Schipper, 1993). The flat

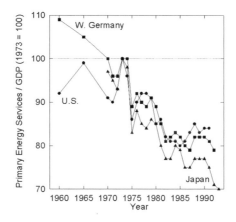

FIGURE 8   Demand for primary energy services in manufacturing, normalized to 1973.

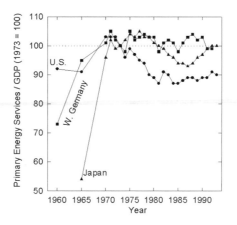

FIGURE 9   Demand for primary energy services for travel, normalized to 1973.

ratio for the United States and Japan suggests that efficiency gains in devices are offsetting the many upward pressures from life-style changes that we have described. Overall, we do not see the consistently declining ratios that would signal an "energy saturation" of the whole society.

A LOOK TO THE FUTURE

What broad speculations and questions may be offered about how future life-styles and policies may affect activities related to energy use in a wide range of areas?

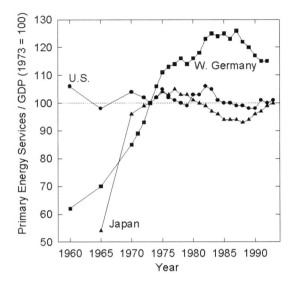

FIGURE 10 Demand for primary energy services for households, normalized to 1973.

Family, Health, Welfare, and Aging

Fertility rates have fallen below replacement in many European countries and in Japan, jeopardizing future social security. New attitudes or policies might encourage more births per adult and the maintenance of larger households in the wealthy nations. At the same time, small households might come to pervade the rest of the world. Household growth could multiply energy use more than population growth.

The aged, ill, and poor might find themselves close to each other in community-care centers or associated with nuclear families. With the former, both the total amount of conditioned space and total mobility could fall. The increased clustering of people expands the possibilities for using collective modes to provide transport or to bring services to the communities. In this view of social security, energy use drops. An alternative is that the aged will remain independent and find themselves increasingly healthy and wealthy, able to travel more, while the ill are better healed and less impoverished, with the poor and homeless made less poor and in a position to find homes. This "ideal" situation, the goal of most politicians, may be elusive for some, but if progress was made towards achieving it, individual comfort and mobility would increase, and with these, energy use as well.

Homes and Goods

Housing size in area per capita might reach a saturation point as people choose to invest marginal income in the quality of their indoor environment, not in quantity. Governments might tighten the direct and indirect subsidies for the provision of housing. Appliance holdings could also saturate, mainly for reasons of space limitations, but at present the array of electronic equipment continues to increase. The number of cars will approach the number of people who can have drivers' licenses, so access to personal mobility may saturate. Yet, the wealthy today show no reluctance to acquire space and goods, and their behavior may simply be standard in a future society with higher incomes.

Utilization of Goods

The utilization of many household goods could likewise reach a saturation point, both because the amount of meat to refrigerate satiates and because the time consumers have to spend using devices is limited. Increasingly, sophisticated refrigerators may be quieter rather than larger, with more varied compartments. Washing and drying equipment can be "smarter" in order to treat the clothes and dishes better, and this tends to reduce energy use for washing and drying. Because the electricity use for electronics is small compared with that for space conditioning, refrigeration, or washing/drying, expanded use of electronics might not significantly raise electricity use. In most developed countries, household energy use appears likely to grow more slowly than does income. Japan could be an important exception, where growth in comfort for heating and cooling offsets improvements in efficiency, particularly as more men spend time at home and families find it more difficult to move about in congested cities and suburbs.

Paid Work

Total work may itself increase or decrease. Will we try to stimulate work in ways that reduce energy use in both manufacturing and service industries? An increased number of hours worked per working family, especially the increased participation of women, appears to have been the main source of increased per capita incomes over the last twenty years. This has led to an increased use of transport for commuting but limited the rise in free time, which is itself transport-intensive. Reduced average weekly working hours could spread employment while also reallocating free time. But fears of future economic insecurity may also stimulate workers to work more, even for lower wages. Will we limit growth in wages to lessen unemployment, thereby giving consumers less disposable income? The main issue is probably whether pressures on those active in the labor force will limit the increase in free time.

Time Use and Location

Child care or help for the aged may keep more parents and older children in their homes. However, collective provision of these services, private or public, may liberate even more hours for work or other activities outside the home. Electronics (pagers, rapid access to assistance) may facilitate monitoring the care of children or the elderly. The low cost of home electronics may attract more people to increase their shopping and free time at home. But the same technologies may liberate individuals from time spent in the home. Computers and "smart homes" are likely to assume a greater role in cooking and other household chores, which will allow occupants to be gone longer while menial tasks are performed in their absence.

Opening hours for services may continue to expand in the United States and will probably be "liberated" where they are now restricted in Europe, as in Britain and Germany. The motivation will be twofold: a better utilization of scarce space in expensive urban areas; and a better utilization of the transport system outside peak hours, which now span from early morning to dinner. With service vehicles and energy-intensive express-delivery systems filling the nightly valley between the morning and evening commutes, much will change.

The cost of mobility and access cannot fail to influence where time is spent. Expensive mobility and restricted access keep people from moving about and will probably increase time spent at home even if other activities there do not change. Should mobility remain inexpensive and access widen, consumers will spend increasing amounts of free time "on the road," somewhere other than at home or at work.

Communications

Communications technologies are permeating the house at low cost, with less wire. Traditional television will yield to electronic program selection from thousands of channels and a vast number of web sites. Wireless phones can substitute for much of the present system. Even if business travel rises, travel for services and leisure could fall as the home becomes a major focus of catalog shopping and the self-production of entertainment services.

Transit

Trends suggest continued growth roughly in proportion to income. Yet, fiscal constraints on road-building, the high real costs of providing underutilized city transit, and the questionable economics of high-speed rail in all but the most well-traveled corridors may force many more travelers to confront the real, high marginal costs of travel (Johnson, 1993). Congestion and noise concerns could lead to congestion pricing and fees to reduce traffic into, through, and within very built-up areas. Air pollution could restrict travel, and becomes a powerful reason in many cities to raise taxes on both dirty fuels and dirty vehicles.[2] Telecommunications could become a widely accepted alternative to travel for shopping, services, and certain businesses and work, although those today with the best and most heavily used communication networks appear to travel more rather than less.

Land Use

Settlement patterns do not appear headed toward the kind of clustering that will produce significant changes in travel time or less energy-intensive modes. Greater residential densities in cities may decrease commuting and travel, or at least permit greater use of collective modes, including walking and biking (New Scientist, 1993). However, settlements will not become denser than they are today without major changes in the way land is valued and taxed. In Europe the suburbs are growing in population, and North Americans stubbornly decline to move toward large city centers, filamenting instead in "edge cities" (Garreau, 1991). Will we tax new land developments to pay for services, possibly discouraging such sprawl?

Leisure Services and Holiday Travel

Several trends point toward an increase in free time for many people. But will they spend it in commercial services (spectator sports, shopping), in nature (walking, exercising), or in long-distance holiday travel? Higher real incomes could mean increasing the "commodification" of free time. Individuals could buy more services at health clubs, more vacations, more entries to museums, concerts, and other events "elsewhere." These demands increase the need for built space and travel and reduce time spent at home. Even "nature" means travel from one's residence in a city or suburb toward the outdoors, increasingly distant from areas developed for homes, services, industry, or agriculture.

In a counter scenario, restricted income growth could mean consumers "do it themselves": gardening, building, and relying on informal, local opportunities for business services and bartered or free services. Environmental concerns could limit both the expansion of leisure developments in "holiday areas" as well as increased access to these areas by hordes of tourists, each trying to get away from the other. Income growth and increasingly cheap travel could contribute to "congestion" of free-time activities and areas, but this development may simply bring new development of leisure and holiday sites throughout the world.

Personal Security

Insecurity, driven by crime, confines people to familiar neighborhoods.

People may stay at home more, using inexpensive electronics for security. Travel would then fall, with its related energy use declining by a far greater amount than the increases for security systems, home comfort, and convenience will demand.

CONCLUDING REFLECTIONS

So far, consumer activities have meant increases in energy use worldwide. In the advanced industrialized nations, such increases have occurred for a long period at a more rapid rate than income growth; however, in recent years, the rate of increase in energy use has grown at a somewhat less rapid rate, as the saturation of certain activities may have taken place. Income-driven life-style changes during the last decades have raised energy use especially for comfort and mobility.

In the formerly planned economies, housing and service-sector comforts will expand greatly as the socialist housing system is replaced. In this case, much of the resulting increase in energy use will offset the one-time savings that appear because of the elimination of earlier senseless overproduction and the circuitous shipping of raw materials. Still, the overall energy intensity of these economies is likely to fall.

In the less-developed countries, income growth should increase demand for all energy services. Improved industrial performance in these countries means greater energy efficiency, but output is likely to increase rapidly. This will allow urban consumers with income to buy increasingly affordable household goods. Indoor comfort increases, and, with growing mechanized mobility, personal mobility will also rise. The reforming and industrializing countries appear to be on much the same track as today's wealthier countries.

In all countries income development matters profoundly; it affects our ability to choose what we in fact do. Other factors hard to forecast also matter. Future levels of mobility may create most uncertainty. Large variations exist even now among relatively similar countries, and it is easy to envision a wide range of travel demand in the future, depending partly on direct costs but at least equally on a host of life-style choices that affect where and how often we move about and why. While a low-mobility future is neither likely nor desirable, it is probable that wealthy countries will take steps to confront users of their transport systems with the real costs of movement and will search aggressively for more environmentally compatible transport systems in the light of likely future growth.

In the absence of information on the efficiency with which primary energy is converted into final energy services to the consumer, it is impossible to conclude that the future will be more or less energy-intensive because of the evolution of consumer life-styles alone. Also, without information on how cleanly energy is generated, we cannot say how life-styles will affect the total environment. Still, we can see from the case of energy that, in the end, consumers and their life-styles arbitrate the quality of the human environment in myriad ways. The driver's license may matter as much as the dynamo.

NOTES

1. Zahavi et al. (1981) claimed a law of constant travel time according to which people on average spend about one hour per day traveling. If true, the distance they travel will increase with speed, particularly when they switch from collective modes to individual cars or motorcycles. The time-budget surveys we have reviewed indicate a slow growth in the time spent traveling. Moreover, vacation travel is almost always excluded from the time-use studies. Because vacation travel appears to be increasing, we surmise that people are spending more time traveling as well as moving faster.

2. The periodic debates in the United States over small gasoline taxes (1–2 cents/mile) suggest that cost-based pricing and incorporation of externalities are not likely to occur soon in that country, but sentiments for such instruments are stronger in Europe.

REFERENCES

Bonke, J. 1986. Aldrig mer Arbejde—oekonomi og verklighed! Copenhagen: Rosinante.
Carlsson, L. G. 1989. Energianvaendning och strukturomvandling i byggnader. R22. Stockholm: Swedish Council for Building Research.
Cooper, R. C., and L. Schipper. 1992. The efficiency of energy use in the USSR—an international perspective. Energy—The International Journal 17(1)1–24.
Diamond, R. C. 1987. Energy Use in Housing for the Elderly: The Effects of Design, Construction, and Occupancy. Berkeley, Calif.: Center for Environmental Design Resources.
Federal Highway Administration. 1982. Nationwide Personal Transportation Study, Household Travel in the United States. Report No. 9. Washington, D.C.: US Department of Transportation.
Federal Highway Administration. 1986. Personal Travel in the United States. Vols. 1 & 2: A Report on Findings from the 1983–1984 Nationwide Personal Transportation Study. Washington, D.C.: US Department of Transportation.
Federal Highway Administration. 1991. Nationwide Personal Transportation Study, Early Results. Washington, D.C.: US Department of Transportation.
Fredbaeck, K. 1979. Fritidshus betydelse foer energianvaendning. Stockholm: Swedish Council for Building Research.
Garreau, J. 1991. Edge City: Life on the New Frontier. New York: Doubleday.
Gershuny, J., and S. Jones. 1987. Time Use in Seven Countries, 1961 to 1984. Bath, England: University of Bath.
Gershuny, J., and I. Miles. 1987. The New Services Economy: The Transformation of Employment in Industrial Societies. London: Frances Pinter.
Gladhart, P., B. Morrison, and J. Zuiches. 1986. Energy and Families. East Lansing, Mich.: Institute for Family and Child Study, Michigan State University Press.
Howarth, R., L. Schipper, and B. Andersson. 1993. The structure and intensity of energy use: Trends in five OECD nations. The Energy Journal 14(2):27–45.
Hu, P., and J. Young. 1993. Nationwide Personal Transportation Study. Washington, D.C.: US Department of Transportation.
Johnson, E. 1993. Avoiding the Collision of Cities and Cars. Chicago: American Academy of Arts and Sciences.
Ketterød, G. H. 1992. Døgnet rundt: Tidsbruk og tidsorganisering 1970–90. Oslo, Norway: Statistik Sentralbyrå.
Meyers, S., L. Schipper, and J. Salay. 1993. Energy use in Poland: Analysis of trends and international comparison. Energy Policy 22:699–713.
Mogensen, G. V. 1990. Time and Consumption: Time Use and Consumption in Denmark in Recent Decades. Copenhagen: Danmarks Statistik.
New Scientist. 1993. July 28.

Robinson, J. P., V. G. Andreyenkov, and V. D. Patrushev. 1988. The Rhythm of Everyday Life: How Soviet and American Citizens Use Time. Boulder, Colo.: Westview Press.

Sanne, C. 1988. Living People. Stockholm: Liber Foerlag.

Schipper, L. 1984. Residential energy use and conservation in Sweden. Energy and Buildings 6(1):15–38.

Schipper, L. 1993. Energy Efficiency and Human Activity: Lessons from the Past, Importance for the Future. Report presented at the World Bank Development Conference, Washington, D.C., May 3–4, 1993.

Schipper, L., S. Meyers, and H. Kelly. 1985. Coming in from the Cold: Energy Efficient Homes in Scandinavia. Report to the German Marshall Fund. Washington, D.C.: Seven Locks Press.

Schipper, L., S. Bartlett, D. Hawk, and E. Vine. 1989. Linking life-styles and energy use: A matter of time? Annual Review of Energy 14:273–320.

Schipper, L., S. Meyers, M. Grubb, M. Chadwick, and L. Kristoferson. 1992a. World Energy: Building a Sustainable Future. Stockholm: Stockholm Environment Institute.

Schipper, L., S. Meyers, R. Howarth, and R. Steiner. 1992b. Energy Efficiency and Human Activity: Past Trends, Future Prospects. Cambridge, England: Cambridge University Press.

Schipper, L., R. Steiner, M. J. Figueroa, and K. Dolan. 1993. Fuel prices, automobile fuel economy, and fuel use for land travel: Preliminary findings from an international comparison. Transport Policy 1(1):6–20.

Sheinbaum, C., and L. Schipper. 1993. Residential sector carbon dioxide emissions in OECD countries, 1973–1989: A comparative analysis. Pp. 255–268 in The Energy Efficiency Challenge for Europe: Proceedings of the ECEEE Summer Study. Vol. II. Oslo, Norway: European Council for an Energy-Efficient Economy.

Szalai, S., ed. 1972. The Use of Time. The Hague: Mouton.

Zahavi, Y., M. J. Beckmann, and T. F. Golob. 1981. The "UMOT"/Urban Interactions. Report No. DOT-RSPA-DBP-10-7. Washington, D.C.: US Department of Transportation.

Technological Trajectories and the Human Environment. 1997.
Pp. 110–134. Washington, DC: National Academy Press.

Elektron: Electrical Systems in Retrospect and Prospect

JESSE H. AUSUBEL AND CESARE MARCHETTI

And I saw something like the color of amber,
like the appearance of fire round about enclosing it;
from what appeared to be his loins upward,
and from what appeared to be his loins downward,
I saw what appeared to be fire,
and there was a brightness round about him.

—Ezekiel 1:27 (circa 595 B.C.)

In the ancient world, *electrum* (Hebrew) or *elektron* (Greek) was the material amber. Amber, when rubbed and electrified, preferably with cat fur, moved and lifted dust specks and small objects. The Greeks first identified electricity by its godlike capacity for *action at a distance.* This capacity and its control have been and will continue to be the trump cards in the invention and diffusion of electric machinery.

While its power and magic are old, electricity as an applied technology is young, with a history of barely more than a century. Two thousand five hundred years passed between Ezekiel and Thomas Edison. Today the electrical system can place power in precise positions in space with an immense range of capacity, from nanowatts to gigawatts. This spatial fingering is made possible by electrical conductors that are immersed in insulating space or solids. The conductors, which are basically metals, are impenetrable to electric fields and can modify and draw them into long thin threads reaching an office, home, or the memory cell in a computer chip.

Electromagnetic waves, as well as wires, transport electrical energy into space. Microwave guides and optical fibers resemble wires fingering into space. Efficient interfaces between the two modes of transport have developed, greatly extending the panoply of gadgets that transform electricity into useful actions.

Electrical technology is one of the few technologies that emerged straight from science and organized research. The lexicon of electricity—ohms, amperes, galvanometers, hertz, volts—is a gallery of great scientists of the eighteenth and nineteenth centuries. Applications of electricity were the subject of the first systematic industrial research laboratory, established in 1876 by Edison in Menlo Park, New Jersey. There, Edison and his colleagues made the phonograph in 1877, a carbon-filament incandescent lamp in 1879, and myriad other inventions.

The earliest attempts to apply electricity came from laboratories studying electrostatic phenomena. Medicine, always curious to test new phenomena in the human body that promised healing or strength, led the way. Many claims sprang from the spark, shock, and sizzle of electrostatic phenomena. Eighteenth-century scientists reported that electric charges made plants grow faster and that electric eels cured gout. They sent electrical charges through chains of patients to conquer disease and, as among the clientele of Dr. James Graham's fertility bed in London, to create life. C. J. M. Barbaroux, later a leader of the Girondist faction in the French Revolution, enthused in 1784:

> O feu subtil, âme du monde,
> Bienfaisante électricité
> Tu remplis l'air, la terre, l'onde,
> Le ciel et son immensité.[1]

Electricity brought to life the subject of Dr. Frankenstein's experiments in Mary Shelley's famous novel, published in 1818. An application of electricity also vitalized the ancient Egyptian in Edgar Allan Poe's 1845 story "Some Words with a Mummy" (Poe, 1976). Upon awakening, the mummy observes to the Americans gathered round him, "I perceive you are yet in the infancy of Galvanism." Later in the nineteenth century the Swedish playwright August Strindberg wrapped himself in currents to elevate his moods and even gave up writing to pursue electrical research until he badly burned his hands in an ill-planned experiment.

Popular imagery notwithstanding, the high-voltage, low-current electrostatic phenomena were at the core of electric research until only about 1800, when Alessandro Volta announced his invention of the battery. Volta introduced the more subtle low-voltage, high-current game of electrodynamics. Twenty-five years linked the flow of electric currents to the force of electric magnets. Another twenty-five years bound the two productively into the electric dynamo and motor.

Among the key figures in the electromechanical game was an American, Joseph Henry, who, with the Englishman Michael Faraday, contributed a series of discoveries leading to practical electric generators. Tracing a bright path back to Benjamin Franklin, electricity was one of the first fields of research in which the United States assumed a leading role, and one of the first technologies to diffuse earliest in America. As we shall see, once the interface between mechanical and electrical power had been invented, the niche for expansion proved immense.[2]

POWER FOR THE WORKSHOP

Since the Middle Ages, water wheels had provided the primary drive for grinding grain, fulling cloth, working metal, and sawing wood. But mechanical power drawn from water or wind did not permit action at a distance, except through even more mechanical devices. These could become sophisticated and baroque. For example, a cable system spread 1 megawatt of mechanical power from the falls of Schaffhausen, Switzerland, to the industrial barracks around them. The mechanically drawn San Francisco cable cars continue to delight visitors but only travel a distance of one or two kilometers.

Powered by water, workshops had to be riparian. "Zavod," the Russian word for a plant, literally means "by the water." Ultimately, steam detached power from place. Over a period of decades, steam engines overtook water wheels. In America, steam needed one hundred years to supersede water. Though we recall the nineteenth century as the age of steam, water did not yield first place until 1870. The primacy of steam in America would then last just fifty years (Figure 1).

At first, steam preserved the layout of the factory. It simply provided more flexible and dependable mechanical energy. The small early steam engines usually operated individual devices. A leap forward came with the advent of the single, efficient, central steam station to serve all the machinery inside a plant. Pulleys rotating above the heads of the workers provided power for their diverse machines via vibrating and clapping belts. But the network of beams, blocks, cords, and drums for transmitting the steam power to the machinery on the floor encumbered, endangered, and clamored.

The electric motor drive, which emerged around 1890, revolutionized the layout of the factory. The first era of electrical systems commenced. The steam engine now ran an electric generator that penetrated the factory with relatively inconspicuous copper wires carrying electricity, which in turn produced mechanical energy at the point of consumption with an electric motor. Here was the seed of modern manufacturing. The electric motor drive permitted the factory machines to be moved along the production sequence, rather than the reverse.

One might suppose that the superior electric transmission, with a generator at one end and motors at each machine, would quickly supplant the old mechanical system. In fact, as Figure 1 shows, the process required fifty years. Resistance was more mental than economic or technical. In 1905 the influential American historian and journalist Henry Adams chose the images of the Virgin and the dynamo around which to write his autobiography (Adams, 1918). The dynamo symbolized the dangerous, inhuman, and mindless acceleration of social change.

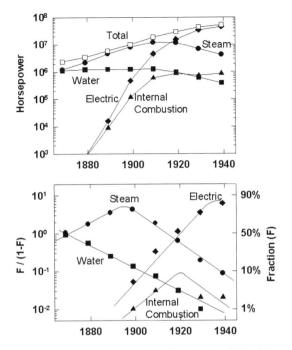

FIGURE 1 Sources of power for mechanical drives in the United States. NOTE: The upper panel shows the absolute horsepower delivered by each type and their sum. The lower panel shows the fraction (F) of the total horsepower provided by each type, according to a logistic substitution model. DATA SOURCE: Devine (1983, Table 3, p. 351).

POWER FOR THE REGION

By the time arcs and lamps emerged from Mr. Edison's workshops, the generator could illuminate as well as grind, cut, and stamp. But the paradigm of the single generator for the single factory was soon superseded by the idea of a generator, or, better yet, a power plant, serving an entire community.

At first, electric companies were necessarily small. Technology for the transport of electricity particularly limited the scale of operations. The original Edison systems were based on low-voltage direct current (dc), which suffered drastic energy losses over distance. Each piece of territory thus required its own company, and founding a new company meant filling a piece of territory or market niche.

Consider eastern Pennsylvania, a coal-and-steel region where some of the earliest Edison utilities began (Figure 2). Entrepreneurs swarmed the area to spread the successful innovation. About 125 power-and-light companies were established between the middle 1880s and early 1920s, with 1897 being the year

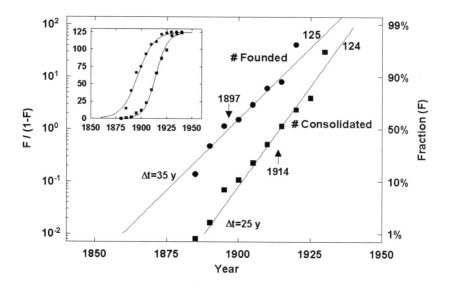

FIGURE 2 Founding and consolidation of electric companies in the United States. NOTE: The main figure presents the two sets of data shown in the inset panel fitted to a linear transform of the logistic curve that normalizes each process to 100 percent, with estimates for the duration of the process, its midpoint, and saturation level indicated. DATA SOURCE: Pennsylvania Power and Light (1940).

of peak corporate fertility. The rush to form companies was a cultural pulse, diffused by imitation.[3]

The evolution of technology to transport electricity, combined with the increase in the density of consumption (kW/km²), made higher transmission voltages economical and progressively coalesced companies. The key technology, first explored in the 1880s by the inventor Nikola Tesla, was alternating current (ac), which could be raised in voltage through transformers and then transmitted long distances with low losses. The merger wave crested in 1914. By 1940 the resulting process left only Pennsylvania Power and Light in operation.

When companies cover a geographical space, their natural tendency is to coalesce, like soap bubbles, especially if a technology permits the larger scale physically and encourages it economically. Several non-technical factors, including government and consumer fears about monopoly, can set limits on scale. Early in the century, Samuel Insull's "electricity empire," centered in Chicago, evoked public regulation, which became normal for the industry. Rapid growth and change usually elicit external regulation. Still, the systems grow in the long run, as we shall see.

In the provision of electric power, the overriding independent variable is *spatial energy consumption*. Its increase leads to higher-capacity transport lines

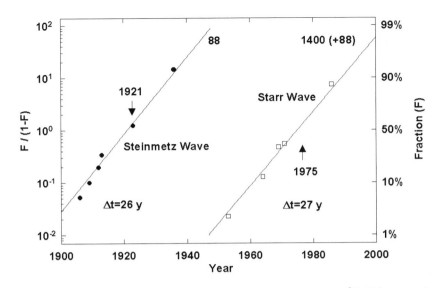

FIGURE 3 Capacity of top US power lines. NOTE: The units are $kV^2/1,000$—a rough measure of power capacity. This figure as well as Figures 4, 6, and 8 show a two-phase process analyzed as a "bi-logistic" normalized with a linear transform. In essence, one S-shaped growth curve surmounts another. The actual values are the sum of the two waves, once the second wave is under way (see Meyer, 1994). DATA SOURCE: Edison Electric Institute, Washington, D.C.

using higher voltage, making it possible to transport energy over longer distances with generators having higher power. This "higher and higher" game led the United States from the 10-kilowatt generator of Edison to the 1-gigawatt generators of today, one hundred thousand times larger.[4]

In fact, the expansion divides into two eras, as we see in Figure 3, which shows the evolution of the maximum line capacity of the US electric system. For the line-capacity indicator, we take over time the square of the highest voltage that is operational. Although various factors lower actual line capacity in practice, this indicator provides a consistent measure of power capacity for analysis of long-term trends.[5] The maximum line capacity grows in two waves, one centered in 1921 and the second fifty-four years later in 1975.

We label the first wave "Steinmetz," for Charles Proteus Steinmetz, the founding leader of the engineering department of the General Electric Company (GE) and a symbol of the fruitful interaction of mathematical physics and electrical technology (Hammond, 1924). Following the pioneering work of Tesla, Steinmetz began investigating the problems of long-distance transmission and high-voltage discharges around 1905. The spectacular success of GE in subsequent decades testifies to the timeliness of Steinmetz's innovations. New alternating-current systems and related gadgets made huge profits for GE and the other leading equipment supplier, Westinghouse, and incidentally killed many small-scale utilities, as in Pennsylvania.

The second pulse of growth in line voltage reaches a temporary ceiling at about 1.5 megavolts. Interestingly, the stretches of innovative activity, as measured by the interval to achieve 10 to 90 percent of the system development, cover only about half the time of electricity's waves of growth. Two to three decades of rapid expansion are digested in a comparably long period of stability and consolidation, a frustrating cycle for engineers. Again the limit may not be technical or economic, but social. Society tailors the expanded system to fit its norms for safety and harmony. One constraint is available rights-of-way, which are very limited at present.

Because the area of the United States is constant and filled by the electrical network, total generating capacity approximates the spatial density of consumption. The growth in installed generating capacity also splits into two pulses, centered around 1923 and 1971 (Figure 4). At peak times operators experience the most rapid change and customers suspect the operators' ability to handle it. During the second wave, annual growth in consumption peaked in the 1950s and 1960s at more than 10 percent per year for many US utilities. The system in the Northeast blacked out one day in November 1965, prompting regional power pooling arrangements. To address concerns about the reliability of the entire network, the industry consorted to form the Electric Power Research Institute,

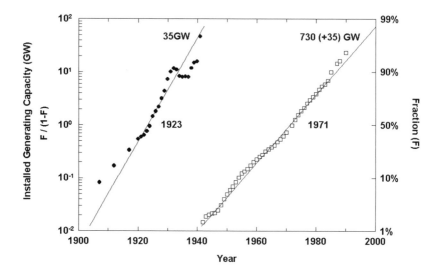

FIGURE 4   Installed electric generating capacity in the United States. DATA SOURCES: US Bureau of the Census (1978a, b; 1981; 1984; 1986; 1989; 1991; 1992; 1994).

which opened its doors in 1973 under the leadership of Chauncey Starr, for whom we name electricity's second wave (Starr, 1995).

The current pulse of growth in US generating capacity reaches a ceiling around 765 gigawatts. The actual system growth has exceeded 90 percent of the niche, which in our view explains the recent slowdown in the building of power plants, nuclear or other, in the United States. The system anticipated the growth in demand that is tuned to economic development and technological diffusion, boxed into the long, roughly fifty-year economic cycles that have characterized the last two hundred years (Marchetti, 1986). At the end of the cycles, demand lags and overcapacity tends to appear.

Will the higher-and-higher game resume? In both line voltage and generating capacity, the growth in the second electrical wave exceeded the first by more than an order of magnitude. If the pattern repeats, the increase in electricity consumption will lead to ultra-high voltage lines (for example, ± 2 megavolts) with higher capacity (for example, 5 or 10 gigawatts) and *continental range*. The great advantage of continental and *intercontinental* connections is that standby reserves and peak capacity can be globalized. The worldwide load would be smoothed over the complete and immanent solar cycle. Generators could also become very large, with according economies of scale.

If the system evolves to continental scale, the much-discussed superconductivity at room temperature might not revolutionize transmission after all. Energy lost in transport and distribution is a stable 10 percent, a huge amount in absolute terms, but too small to change the basic economics if 2-megavolt lines cover the continents. Superconductivity could, however, bring about a revolutionary drop in the size of machinery, thereby permitting the construction of units of larger capacity.

Continental scale surely means increased international trade in electricity. All territory looks the same to electricity. If available technology is employed, electricity will stream across borders despite the political barriers that typically impede the easy flow of goods and ideas. Today Europe exchanges electricity almost freely. Italy buys from France the equivalent production of six 1-gigawatt nuclear reactors either via direct high-voltage lines or through Switzerland. Electricity trade could form a significant component of international payments over the next fifty to one hundred years, requiring reorganization and joint international ownership of the generating capacity. Electricity trade between Canada and the northeastern United States already elicits attention.

UTILIZATION AND CAPACITY

The utilization factor of generation plants counts heavily in the economy of the system and indicates the quality of its organization. The US electric industry searched successfully between 1910 and 1940 for efficient organization, notwithstanding the Great Crash of 1929, as the average annual utilization climbed from two thousand to above four thousand hours, a utilization rate of about 50 percent (Figure 5). The rise owed to spatial integration and the reduction of reserves consequent to the introduction of high-capacity transport lines with increasing operating voltage as well as the coordination of network dispatch to use plants more effectively.

Since 1940 the system appears to have fluctuated around a utilization rate of 50 percent. Generators with low capital cost and high variable cost combine with base-loads plants with high capital cost and low variable cost to determine the current usage level. Although the utilization factor surely has a logical upper limit quite below 100 percent, even with high-voltage lines having continental reach, a 50-percent national average appears low, notwithstanding scorching August afternoons that demand extra peak capacity.

Breaking the 50-percent barrier must be a top priority for the next era of the industry. Otherwise, immense capital sits on its hands. One attractive way to make electric capital work around the clock would be to use plants at night. The mismatched timing of energy supply and demand existed when water power dominated. Pricing, automation, and other factors might encourage many power-consuming activities, such as electric steel-making, to go on the night shift. Nuclear heat, generating electricity by day, could of course help to make hydrogen at night. The ability to store hydrogen would make the night shift productive.

The nearness of overcapacity in the electrical system also creates suspicion that forecasting within the sector has not been reliable. Analyses of projections of total electricity use made by the US Department of Energy and others fuel the suspicion. Reflecting a period when electricity consumption had doubled in spans

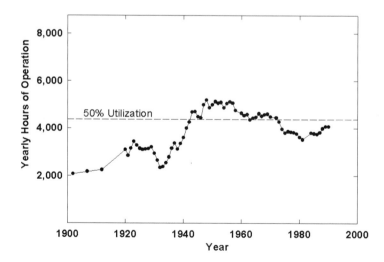

FIGURE 5   The rate of utilization of US electric generating plants.   DATA SOURCE: US Bureau of the Census (1978a, b; 1981; 1984; 1986; 1989; 1991; 1992; 1994).

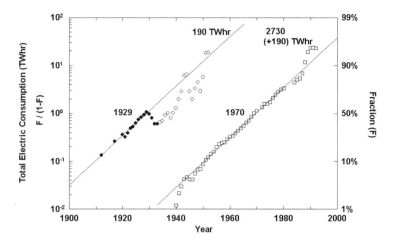

FIGURE 6 Total US electric consumption. NOTE: Here and in Figure 8 the empty circles indicate periods of overlap in the sequential growth waves. Assigning the exact values to each wave during the periods of overlap is somewhat arbitrary. DATA SOURCE: US Bureau of the Census (1978a, b; 1981; 1984; 1986; 1989; 1991; 1992; 1994).

of ten years, in 1978 federal officials projected an increase by 1990 from 2,124 terawatt hours to 4,142 terawatt hours.[6] The actual level for 1990 was 2,807 terawatt hours.

Can we do better? Fitting the data for total utility electric use to our model with data through 1977 yields an estimated level of about 2,920 terawatt hours for the growth pulse now ending (Figure 6). Net generation in 1993 was 2,883 terawatt hours. Projecting electricity demand matters because it influences investments in capacity. Accurate projections might have lessened the pain for the utilities, which ordered and then canceled plants; the equipment suppliers, who lost the orders; and consumers, who ultimately pay for all the mistakes.

POWER FOR THE HOME

As suggested earlier, electricity is a spatial technology. Conquering a territory means connecting with potential users. We tend to think that almost everyone was connected soon after the first bulb was lit, but in fact the process extended gradually over fifty years and culminated even in the United States only in mid-century (Figure 7). Although slowed by the Great Depression, non-rural hookups reached 90 percent of the market by 1940. Rural areas joined the grid about one generation later than cities, reaching a midpoint of the process in 1943 versus 1920 for the townsfolk. This interval measures the clout of rural politi-

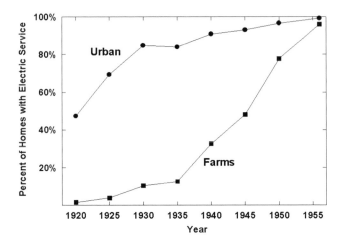

FIGURE 7 Percentage of US homes with electric service. DATA SOURCE: US Bureau of the Census (1978a).

cians, who secured subsidies for the costly extension of power lines to areas of low population density, as well as the conservatism of the countryside.

The data further confirm that electricity's first century has encompassed two eras. During the developmental spread of the system until about 1940, most electricity went for industry and light, substituting for other energy carriers in already existing market niches. In the second era, electricity powered new devices, many of which could not have performed without it, such as televisions and computers. Most of the new demand came in the residential and commercial sectors.

Average residential consumption has increased by a factor of ten since 1940 and appears in our analyses to saturate in the 1990s at about 10,000 kilowatt hours per year. One might say that the customer is the home, not the human. Home appliances have increased by the tens and hundreds of millions: refrigerators, video-cassette recorders, vacuum cleaners, toasters and ovens, clothes washers and dryers, dishwashers, air conditioners, space heaters, and, more recently, personal computers, printers, and fax machines.

We emphasize the residential because it is becoming the number-one consumer. Residential consumption has grown faster than other major sectors over the past decades and in 1993 overtook industrial consumption in the United States. The number of housing units has grown sevenfold in the United States since 1900, while the number of people has tripled, as residents per unit have declined and second homes increased (see Schipper, this volume). As the second wave of electrification reaches its culmination, the residential share appears destined to plateau at about 35 percent of the total use of electricity, more than twice its share of the first wave. In a third wave of electricity, residential consumption may grow only at the same rate as overall consumption, or, if life-styles continue to include more home space and reduced working time, at an even faster rate (Ausubel and Grübler, 1995). Californians already spend more than 60 percent of all their time at home indoors (Jenkins et al., 1992). So do New Yorkers and Indians.

CLEANING THE HUMAN ENVIRONMENT

In the absence of electricity, we heat, light, and power our homes and workplaces with wood, coal, kerosene, oil, manufactured city gas, and lesser fuels. Electrification has thus meant a cleaner, safer, and healthier environment at the level of the end-user, once protections against shock and other hazards were properly wired into the system. Dangers associated with open fires and smoke diminished. Better-ventilated homes and workplaces lessened exposure to influenza, pneumonia, tuberculosis, diphtheria, measles, whooping cough, scarlet fever, and other airborne threats. Modern refrigeration in homes, shops, trucks, and railroad boxcars reduced the numerous waterborne gastrointestinal threats.

Environmentally, electricity concentrates pollution at a few points. At these points we can deal with the problems or not. The main question then becomes: What is the primary energy source for the generation? The most wanted environmental culprit is carbon, and so the main environmental challenge for electricity may be summarized by the measure of the carbon intensity of electricity production, for example, the ratio of carbon by weight to kilowatt hours generated.[7] In the United States, this ratio fell by half between 1920 and 1940, from about 500 metric tons of carbon per gigawatt hour produced to about 250. Since the 1940s, the US ratio has fallen below only about 200 metric tons per gigawatt hour and has remained rather flat in recent decades because coal has gained markets in electric power plants, offsetting efficiency gains in the operations of the plants as well as gains in terms of reductions that oil and especially gas would have contributed. Many other countries have continued to create more watts with fewer carbon molecules. The powerful underlying evolution of the energy system from coal to oil to natural gas to nuclear or other carbon-free primary sources will bring reductions (Ausubel, 1991). The world appears a bit past the middle point of a decarbonization process that will take another 150 years for completion. The United States will not long remain apart from the global movement.

Electricity production was originally based on coal alone. At present, it is the only outlet for coal. Even steel-making, which historically consumed a substantial fraction of coal (sometimes more than 10 percent), abandoned coal, dropping demand. Coal will fight hard to keep its last customer. Interestingly, electricity was never linked to oil, one of the other major transforming technologies of the twentieth century. Electricity and oil may now begin to compete seriously for the transport market, as we discuss later. Natural gas is already penetrating the electrical system thanks to the great flexibility, low capital cost, quick starts, and efficiency of gas turbines. At present, electricity remains the only product of the nuclear system. Approaching an energy system with zero emissions, about which all environmentalists dream, will require nuclear to diversify into the hydrogen-making business. The team of electricity and hydrogen can eventually solve all the problems of pollution at the level of the end-user of energy.

Electrical systems can add visual pollution with their network of towers, wires, and poles. Militant Greens already dynamite pylons and will accept no new structures. New technologies can increase the capacity of the existing lines and diminish intrusions. In this regard, direct current, now ultra-high, may have a second life as a bulk carrier aided by relatively cheap electronics, such as thyristors, which are capable of transforming all types of units of electricity into all others. Burying power lines might beautify the landscape, as well as lessen fears about the health effects of electromagnetic fields.

FEEDING THE ELECTRICAL SYSTEM

A growing share of primary fuels generates electricity; again, two waves are evident (Figure 8). At the world level, the first centered in 1920 and the second in 1970. The present wave is saturating at close to 40 percent. For the United States, the current wave appears to have saturated at about the same level.

Is there a limit to the fraction of fuels feeding into the electrical system?

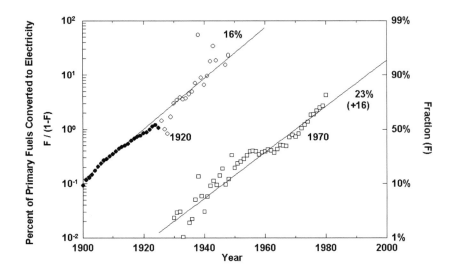

FIGURE 8 Percentage of world primary fuels converted to electricity. DATA SOURCE: Nebojša Nakićenović, personal communication, 1995.

Many energy buffs postulate a ceiling at around 50 percent. A third era of electrical growth does seem likely to occur. Electricity is more flexible and fungible than hydrocarbon fuels. The innumerable devices of the information revolution require electrical power. The transport sector, which has remained largely reliant on oil, could accept more electricity. But the drawbacks are the inefficiencies and the costs of the transformation.

Inefficiencies are eventually eaten up (Nakićenović et al., 1990). A successful society is, after all, a learning system (Marchetti, 1980). In fact, perhaps the greatest contribution of the West during the past three hundred years has been the zeal with which it has systematized the learning process itself through the invention and fostering of modern science, institutions for retention and transmission of knowledge, and diffusion of research and development throughout the economic system. But learning may still go slowly when problems are hard.

THE SIX-HUNDRED-YEAR WAR FOR EFFICIENCY

The degree of difficulty for society to learn about power and light shows quantitatively in the duration of the process improvements illustrated in Figure 9. Technologists fought for three hundred years to bring the efficiency of steam power generation from 1 percent in 1700 to about 50 percent of its apparent limit today. Electrical energy is glorified as the purest form of free energy. In fact, the heat value of other fuels when they burn also corresponds to free energy. Thus, the thermodynamic limit of electric generators is 100 percent. Of course, it can be very difficult to reduce losses in combustion. Still, we may muse that during the next three hundred years efficiency will go to 99 percent.[8] This long trajectory suggests that the structure upstream for power generation does not leave much room for spectacular breakthroughs.

Still, 70-percent efficiency can be eyed as the next target, to be achieved over fifty years or so. Turbosteam plants with an efficiency of about 60 percent have been constructed. Although further gains in this regard appear limited, the massive diffusion of highly efficient turbine technology is sure to be a lucrative and influential feature of the next fifty years or so. Fuel cells, avoiding the free energy loss almost inevitable in the combustion process on which turbines rely, may well lead to the even higher efficiencies. Electrochemistry promises such technology but mentally seems more or less still stuck in Edison's time. Perhaps solid-state physics can produce the insights leading to the needed leap forward as specialists in this field become more interested in surfaces, where the breakthroughs need to occur.

At the 70-percent level of efficiency, an almost all-electric distribution of primary energy looks most appealing. The catch is the load curve, which seems likely to remain linked to our circadian rhythms. In Cole Porter's song lyric, we hear "Night and day, you are the one"; but in energy systems night still dims demand and means expensive machinery remains idle. Even in cities famous for

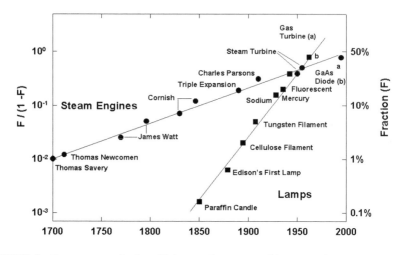

FIGURE 9   Improvement in the efficiency of motors and lamps analyzed as a sigmoid (logistic) growth process. NOTE: Shown in a linear transform that normalizes the ceiling of each process to 100 percent. MAIN DATA SOURCES: for lamps, *Encyclopaedia Britannica* (1964); for motors, Thirring (1958).

their nightlife, nocturnal energy demand is only one-third of the daytime requirement. The ratio of day to night activity does not seem to have changed much. The ancients actually spent considerable time awake at night, despite miserable illumination. The fine old word "elucubrate" means to work by the light of the midnight oil, according to the *Oxford English Dictionary*.

Even if most humans continue to sleep at night, we have pointed out earlier that their energy-consuming machines can work nocturnally. In fact, remote control and the shrinking work force required to operate heavy industry ease the problem. So, too, will linking parts of the globe in sun and shade, summer and winter.

Still, we should clearly look further for efficiency gains. Much large electrical machinery is already so efficient that little or no gain is to be expected there. But a discontinuous step could yet come in the progress of machinery. Superconductivity, when it permits high magnetic fields, can lead to compactly designed motors with broad applications and very low energy losses. The proliferation of numerous micro-machines will of course tend to raise electricity demand, partially offsetting the efficiency gains they offer. The miniaturization of circuits and other aspects of computing systems in the past two decades shows how powerfully reducing the size of objects can increase their applications and numbers.

The Splicer

In proposing a more general solution we need to introduce another consideration, namely, reliability. The main drawback of an electrical system is that it permeates the web of social services, so that a breakdown, even for a few hours, can bring tragedy. A defense against this vulnerability, as well as a means of addressing cyclical loads, could come with the diffusion of multipurpose mini-generators at the level of individual consumers. In effect, we would delegate base load to the global system, leaving peaking and standby to a new multipurpose household appliance. Multipurpose means the device could produce heat, electricity, and cold on demand.

Such combined thermal, electric, and cooling systems, which we will call "splicers," are under development. Attempts so far, such as the FIAT TOTEM, have been unsuccessful, in part because the marketed models lack the basic characteristic of *zero maintenance* required by household gadgets. Still, the scheme is appealing, both functionally and economically. The Japanese are doing a sizable amount of research and development in what appears to be a promising direction: stirling engines with free-floating pistons and a power output of a few kilowatts. The machines are maintenance-free, silent, and can compress fluids for the heating and cooling cycles on top of producing electricity with linear oscillating generators. The models described in the literature are powered by natural gas.

In conjunction with a clean gas distribution system, the penetration of the splicer as a home appliance over the next fifty years could revolutionize the organization of the electrical system. The central control could become the switchboard of millions of tiny generators of perhaps 5 kilowatts. Electric utilities might initially abhor the technology that brings such functional change, but already some plan to use it. One attraction is that the final user immediately pays the capital cost.

In any case, the breakthroughs may come instead on the side of the consumers. A number of well-known machines and appliances need technological rejuvenation, as efficiencies are systematically low. And new machines need to be invented. At a high level of abstraction, human needs are invariant: food, clothing, shelter, social rank, mobility, and communication (a form of mobility where symbols move instead of persons or objects). Let us guess the shape of the new machines in the areas of vision and warmth.

Efficient Vision

Illumination, the first brilliant success of electricity beyond powering the workshop, provides a good example. Breaking the rule of the night is an old magical dream. The traditional tools—oil lamps, torches, and candles—were based on a flame with relatively low temperature and small amounts of incandescent soot to emit the light. They performed the task poorly (see Figure 9).[9] The typical power of a candle is 100 watts, but the light efficiency is less than 0.1 percent.

Electricity fulfilled the dream, almost from the beginning, with arc lights, whose emitting source was solid carbon at temperatures of thousands of degrees centigrade.[10] The light was as white as the sun, and efficiency reached about 10 percent. The technical jump was enormous. Theaters, malls, and monuments were lavishly illuminated. People were seduced by the magic. Amusement parks such as Luna Park and Dreamland at Coney Island in New York drew millions of paying visitors to admire the architectural sculptures of light.

Edison's 1879 incandescent lamp was a trifle inferior to the arc in light quality and efficiency but was immensely more practical. Symbolically, in 1882 the New York Stock Exchange installed three large "electro-liers," new chandeliers with sixty-six electric lamps each, above the main trading floor. The exhibition of the power to break the night came first and dramatically. Penetration of the technology came later and, as usual, slowly. US cities, as shown earlier, achieved full illumination only about 1940.

The period from 1940 to 1995 can be called a period of consolidated light. Lamps became brighter and efficiency rose. To the human eye, the quality of the light may actually have worsened with the spread of fluorescents. With laser light, which has terrible visual quality now, we may approach theoretical efficiency, though actual lasers remain inefficient. Will that be the light at the end of the tunnel?

To return to basics, we illuminate in order to see in the dark. Illumination has no value if nobody looks. Arriving in a town at night, we always see the roads brightly lit and empty, so we know of waste. The marvels of the 1980s, electronic sensors and computer chips, can already scan rooms and streets and switch the lights off if no one is present. The watt-watch can help, but we can go further.

Sophisticated weapons systems—those mounted in helicopters, for example—feel the thumb of the pilot, observe his eyes, and shoot where he looks. A camera-computer in a room can watch the eyes of people present and illuminate only what they watch. Phased arrays, familiar in sonars and radars and developed now for infrared emitters, are certainly transportable into the visible range and can create sets of beams that are each directed to a chosen point or following a calculated track. The apparatus might now look baroque, but with miniaturization it could be concealed in a disk hanging from the ceiling of a room. Such a gadget appears to be the supreme fulfillment, illuminating an object only if a human gazes upon it.

But recall again that the objective is not to illuminate but to see. We illuminate because the eye has a lower limit of light sensitivity and, in any case, operating near such a limit is unpleasant. The military has developed complicated gadgets by which scanty photons from a poorly illuminated target are multiplied electronically to produce an image of sufficient luminosity. The principle is good; the machine is primitive. If photons flowing in an energized medium (such as an excited laser crystal) multiplied in a cascade along the way while keeping frequency and direction, we would have invented *nightglasses,* the mirror image of sunglasses.[11] We could throw away all sorts of illuminating devices. A few milliwatts of power would be enough to brighten the night.

Efficient Warmth

The largest part of energy consumed in the home is used for temperature control. Space heating accounts for 60 percent or more of total residential energy use in many developed countries. Heating a home is a notably inelegant process from a thermodynamic point of view. We use pure free energy (electricity or fossil fuels) to compensate for a flow of energy from inside to outside having an efficiency according to the Second Law of Thermodynamics of about 3 percent if the difference in temperature is 10°C. Heat pumps solve the problem conceptually, but they see temperatures *inside* their heat exchangers and consequently overwork.[12] Moreover, operating on electricity generated upstream, they already invite inefficiency into the endeavor.

Consider a radically different proposal. Windows are the big leaks, even when the glazing is sophisticated and expensive. Why not use window panes as *thermoelectric devices,* not to carry heat uphill but to stop heat from sledding downhill, that is, as *heat-flux stopping devices?*

Thermoelectric generators are usually seen as machines to make electricity by using the principle of the thermocouple. However, the device is reversible: by passing electricity through the machine, heat can be moved uphill. Several decades ago refrigerators were proposed using this principle on the basis of its great simplicity, although efficiencies are low. The old scheme for refrigerators could be revised in view of new thermoelectric materials and given suitably competitive objectives.

The basic idea is that electrodes on the inner and outer surfaces of the windowpanes can be made of conductive, transparent glasses. Glass made of zinc oxide might be sufficiently conductive. Voltages across the glass would be very low—volts or fractions of volts. Holding a temperature differential with zero flux would be more efficient energetically than putting heat (electrically!) into a house to balance the outgoing flux.

Electric Motion

So far we have looked at examples where efficiency wins, and net demand for power grows, only if the human population and its use of devices increase faster than efficiency. Now let us look at one example where a large new market might emerge, matching the ultra-high voltage lines and continental connections.

Toward the end of the last century electric motors for vehicle engines attracted much inventive action. Edison and Ferdinand Porsche produced sophisticated prototypes. The idea flopped on the roads but succeeded on the rails. Electric trams clamored through American and European cities, helped create suburbs, and in some cases connected cities. After 1940 most of the system was rapidly dismantled, largely because the trams could not match buses and cars in flexibility or speed. The mean velocity of transport keeps increasing through the progressive substitution of old technologies with new, faster ones. For France, the increase in the average speed of all machine transport has been about 3 percent per year during the last two centuries. Urban and suburban railways have a mean speed of only about 25 kilometers per hour, including stops. Cars have a mean speed on short distance trips of about 40 kilometers per hour. The latest in the series are airplanes, with a mean speed of 600 kilometers per hour. Airplanes will provide most of the increase in mean speed over the next fifty years.

Electric trains succeeded in Europe and Japan for the densely trafficked lines and still operate today. They have decent acceleration and speed compared with diesels. But most trains are not fast; the inclusive travel time on intercity rail journeys is only about 60 kilometers per hour. The fastest trains, the French *trains à grande vitesse* (TGVs), are electric. The question for trains is how to compete with cars on one side and with airplanes on the other. Electricity probably cannot compete with hydrogen for propulsion of cars and other light vehicles.

The great market challenge for the current generation of fast trains, with top speeds of 400 kilometers per hour, is the short distances of less than 100 kilometers along which cars congest and airplanes cannot compete. The present configuration of airports and airplanes are high-speed but low-flux machines. TGVs could prove extremely competitive in the intense shuffling of commuters and shoppers within these distances. A cursory review of Europe reveals about 5,000 kilometers of intercity links fitting the constraints of a 100-kilometer distance and high potential passenger flux.

Fast trains consume more or less the same amount of primary energy per seat-kilometer as a turboprop plane[13] or a compact car. From the power point of view, a running TGV absorbs about 10 kilowatts per seat. The mean power demand of the proposed 5,000-kilometer system of TGV trains for commuters and shoppers would be around 6 gigawatts, with a peak of probably 10 gigawatts. If the concept is successful, this form of transport will be an important consumer of electricity, but it will take at least fifty years to become fully implemented.

To go to very high passenger fluxes over longer distances, one would need to go to aerial configurations of which even the most daring air-transport planners do not chance to dream: flocks of airplanes of five thousand passengers each taking off and landing together like migrating birds.

For intense connections linking large cities with peak fluxes around ten thousand passengers per hour, a solution is emerging that matches system requirements: the magnetically levitated (maglev) train operating in a partially evacuated tube or tunnel. In fact, Swiss engineers have developed the concept of a vacuum version of maglevs in part to reduce drastically the tunnel boring expenses, which in Switzerland would account for at least 90 percent of the cost in a conventional layout (Nieth et al., 1991). To handle the shock wave from a high-speed train, a tunnel normally needs a cross section about ten times that of the train. In addition to narrowing greatly the tunneling requirement, the partial vacuum greatly reduces friction, making speed cheap and thus expanding the operational range of the train.

When operated at constant acceleration—for example, 5 meters per second or 0.5 g (the force of gravity), about what one experiences in a Ferrari sports car—maglevs could link any pair of cities up to 2,000 kilometers apart in fewer than twenty minutes. Consequently, daily commuting and shopping become feasible. Such daily trips account for 90 percent of all travel and are controlled by the total human time budget for travel of about one hour per day. With fast, short trips cities can coalesce in functional clusters of continental size. City pairs spaced less than 500 kilometers or ten minutes apart by maglevs, such as Bonn-Berlin, Milan-Rome, Tokyo-Osaka, and New York-Washington, would especially benefit.

Part of the energy consumption of vacuum maglevs overcomes residual friction; an economic balance must be struck between the friction losses and the pumping power to keep the vacuum. Part regenerates the electromagnetic system that pushes and pulls the trains.[14] The power per passenger could roughly correspond to that of a large car, although these trains may travel at a mean speed of 3,000 kilometers per hour.

The great advantage of the constant acceleration configuration for maglevs is that the energy required for each length of track is constant and could be stored, perhaps magnetically, in the track itself. Power demand is proportional to train speed and moves into the gigawatt range in the central section; however, with local storage (a few kilowatt hours per meter) the external electric power networks would see only the need to make up losses. Even assuming 90-percent efficiency, these would not be negligible. One hundred trains per hour would demand 1 gigawatt for the single line on which they operated.[15] The Swiss system has a final potential of five hundred trains per hour, which would require 5 gigawatts—about one-third of current installed Swiss generating capacity.

The first long-distance maglev will probably run in about five to ten years. Berlin-Hamburg is under construction. The penetration of the technology will be gradual, as major infrastructural technologies always are. In fact, the next fifty years will probably be used largely to establish the feasibility, chart the maglev map, and prepare for the big push in the second half of the twenty-first century. In the long run, maglevs may establish several thousand kilometers of lines and become one of the most important users of electricity. A maglev trip per day becomes a few thousand kilowatt hours per year per person. If India and Eastern China join life in this superfast lane, the picture of a globally integrated, high-capacity electrical system begins to cohere.

CONCLUSIONS

The long economic cycles that seem to affect all parts of social and economic life constitute a good frame of reference for the development of the electrical system in terms of technology, territorial penetration, birth and death of enterprises, and intensity of use. Our examples suggest this is true for the United States and globally.

Two waves of electrification have passed through our societies. In the first, the United States attained system saturation in the 1930s at about 1,000 kilowatt hours annual consumption per residential customer, 200 gigawatt hours of total annual use, 40 gigawatts of installed capacity, and 20 percent of primary fuels producing electricity. In the second wave, we have reached 10,000 kilowatt hours per residential customer, 3,000 gigawatt hours of total use, 800 gigawatts of installed capacity, and about 40 percent of fuels producing electricity.

The fact that the patterns of temporal diffusion and growth are followed makes it possible to fit dynamic equations to the time series of facts and then compare them for consistency. This operation indicates that the 1990s are the season of saturation, which includes the experience of overcapacity or, alternately, underconsumption. Such phases are not uncommon for various branches of the industrial system, as managers tend to assume that growth characteristics of boom periods will extend into recessions, while consumers cut corners.

In the short term, total energy and electric energy consumption may continue to grow at a slower rate than overall economic activity. One interpretation is that during the expansion period of the long cycles the objective is growth, while during the recessive period the objective is to compete, shaving costs here and there and streamlining production. The savings include energy. Meeting goals pertaining to environmental quality and safety further tighten the system.

A new cycle formally beginning in 1995 started the game again, although the effects of the restart will not be particularly visible for a few years. Minima are flat. Looking at the cycles from a distance to grasp the general features, one sees the periods around their ends as revolutionary, that is, periods of reorganization—political, social, industrial, and institutional. We are evidently at this conjunction, and the electrical system will not escape it.

When the electrical system served the village, a complete vertical integration was inevitable. Regional coverage, the preferred scale of the past fifty years, also favored such integration. With the expansion to continental dimensions, a shift in responsibilities may make the system more efficient, agile, and manageable. The typical division is production, trunk-line transport, and retailing, with different organizations taking care of the pieces and the market joining them. The experiments in this sense now running in Great Britain, Australia, and other countries can be used as a test bed to develop the winning ideas.[16]

Apart from various economic advantages and organizational complications, the use of splicers on a large scale—untried to date—may bring an almost absolute resiliency, as every subset of the system may become self-sufficient, if temporarily. The electrical system should also become cleaner, as it intertwines more closely with natural gas and probably nuclear energy, thus furthering decarbonization. A sequence of technical barriers will appear, and thus the process of systematic research and innovation will continue to be needed; it will produce timely results.

In fact, our analyses suggest that rates of growth of technology tend to be self-consistent more than bound to population dynamics. Population, however, defines the size of the niche in the final instance. Thus a key question is, how long will it take to diffuse Western electric gadgetry to the 90 percent of the world that is not already imbued with it? The gadgetry keeps increasing. Followers keep following, if more closely. Based on historical experience, diffusion to distant corners requires fifty to one hundred years. Even within America or Europe, as we have seen, pervasive diffusion takes that long for major technologies. So most people may have to wait for most of the next century to experience nightglasses, splicers, and maglevs. These devices may be largely features of a fourth wave of electrification, while the spread of the profusion of information-handling devices dominates the third wave that is now beginning.

Considered over centuries and millennia, the electrical adventure is deeper than a quest for gadgets. In 1794 Volta demonstrated that the electric force observed by Luigi Galvani in twitching frog legs was not connected with living creatures, but could be obtained whenever two different metals are placed in a conducting fluid. Today we use electricity to dissolve the difference between inanimate and living objects and to control and inspire the inanimate with more delicacy than Dr. Frankenstein. Introducing electricity into production raised the rank of workers from sweating robots to robot controllers. The process can be generalized, with humanity—at leisure or at work—giving orders to its machines by voice or a wink of the eye. This ancient aspiration for action at a distance and direct command over the inanimate will drive invention, innovation, and diffusion for hundreds of years more; we come full circle to the *elektron* of the ancient Hebrews and Greeks.

ACKNOWLEDGMENTS

We thank Perrin Meyer, for research assistance and figure preparation, as well as Arnulf Grübler, John Helm, Eduard Loeser, Nebojša Nakićenović, and Chauncey Starr.

NOTES

1. "Oh subtle fire, soul of the world, / beneficent electricity / You fill the air, the earth, the sea, / The sky and its immensity." Quoted in Darnton (1968, p. 29).

2. For general histories of electrification, see Hirsch (1989), Hughes (1983), Nye (1990),

Schivelbusch (1988), and Shurr et al. (1990). For data and information on the early history of energy and electricity, see Schilling and Hildebrandt (1977).

3. Such diffusive processes are well fit by the logistic equation, which represents simply and effectively the path of a population growing to a limit that is some function of the population itself. For discussion of applications of logistics, see Nakićenović and Grübler (1991). On the basic model, see Kingsland (1982).

4. A kilowatt (kW) is 1,000 watts; a megawatt (MW) is 1,000,000 W; a gigawatt (GW) is 1,000 MW; a terawatt (TW) is 1,000 GW. US generating capacity was 735 GW in 1990.

5. Power is equal to V^2/R, where V is voltage and R is resistance.

6. For an analysis of electricity projections, see Nelson et al. (1989).

7. Sulfur and other emissions from power plants also cause ills, but these have proven to be largely tractable (see Nakićenović, this volume).

8. While Carnot efficiency (now about 60 percent) limits heat cycles, fuel cells do not face such a limitation, as they are not based on heat cycles.

9. Gaslight, with a mantle with rare-earth elements, was a superior source of bright light for a period.

10. The plasma struck between the two carbon electrodes also emits.

11. Sticking to monochromatic light, a ray proceeding in a resonantly excited medium stimulates emission and becomes amplified. Amplification is relatively small with present devices; hence the ray must travel up and down between mirrors. But no physical law limits amplification to such low levels. Semiconductor lasers, pumped by electric voltage, might hold the solution. In a second stage, they should also operate for a number of colors.

12. The equivalent free energy of heat flowing out of a building is measured through the temperatures inside (T_1) and outside (T_2) in kelvin and is (T_1-T_2)/T_1. In the case of a heat pump, due to temperature drops in the heat exchanger, it pumps heat from a temperature lower than T_2 into a temperature higher than T_1.

13. For example, airplanes of the type ATR-42 or Dash.

14. We can calculate the amount of energy circulating in the system for a maglev with constant acceleration operating over a distance of 500 kilometers. A train of 200 metric tons accelerating at 0.5 g has a pull force (drag) of 1,000 kilonewtons, which over a stretch of 500 kilometers corresponds to 5×10^{11} joules, or approximately 140,000 kilowatt hours. A mean loss of 10 percent would require 14,000 kWh for one thousand seats, or 14 kWh per seat over 500 km. This would correspond to 84 kW per passenger at a typical trip time of 10 minutes (e.g., Bonn to Berlin).

15. For example, fifty trains might operate in each direction, spaced one minute apart. They can start from different stations or lanes. One per minute would be the frequency in the neck of the tunnel.

16. For discussions of models of organizational change in the electricity industry, see Sack (1995, 1996).

REFERENCES

Adams, H. 1918. The Education of Henry Adams: An Autobiography. Boston: Massachusetts Historical Society. Reprinted in 1961 by Houghton Mifflin, Boston.

Ausubel, J. H. 1991. Energy and environment: The light path. Energy Systems and Policy 15(3):181–188.

Ausubel, J. H., and A. Grübler. 1995. Working less and living longer: Long-term trends in working time and time budgets. Technological Forecasting and Social Change 50(3):195–213.

Darnton, R. 1968. Mesmerism and the End of the Enlightenment in France. Cambridge, Mass.: Harvard University Press.

Devine, W. D. Jr. 1983. From shafts to wires: Historical perspective on electrification. Journal of Economic History 43:347–372.

Encyclopaedia Britannica. 1964. Chicago: Encyclopaedia Britannica.

Hammond, J. W. 1924. Charles Proteus Steinmetz: A Biography. New York: The Century.

Hirsch, R. F. 1989. Technology and Transformation in the American Electric Utility Industry. Cambridge, England: Cambridge University Press.

Hughes, T. P. 1983. Networks of Power: Electrification in Western Society. Baltimore: Johns Hopkins University Press.

Jenkins, P. L., T. J. Phillips, E. J. Mulberg, and S. P. Hui. 1992. Activity patterns of Californians: Use of and proximity to indoor pollutant sources. Atmospheric Environment 26A(12):2141–2148.

Kingsland, S. 1982. The refractory model: The logistic curve and the history of population ecology. Quarterly Review of Biology 57:29–52.

Marchetti, C. 1980. Society as a learning system. Technological Forecasting and Social Change 18:267–282.

Marchetti, C. 1986. Fifty-year pulsation in human affairs: Analysis of some physical indicators. Futures 17(3):376–388.

Meyer, P. S. 1994. Bi-logistic growth. Technological Forecasting and Social Change 47:89–102.

Nakićenović, N., and A. Grübler, eds. 1991. Diffusion of Technology and Social Behavior. Berlin: Springer-Verlag.

Nakićenović, N., L. Bodda, A. Grübler, and P.-V. Gilli. 1990. Technological Progress, Structural Change and Efficient Energy Use: Trends Worldwide and in Austria. International Part. Laxenburg, Austria: International Institute for Applied Systems Analysis.

Nelson, C. R., S. C. Peck, and R. G. Uhler. 1989. The NERC fan in retrospect and prospect. The Energy Journal 10(2):91–107.

Nieth, R., W. Benoit, F. Descoeudres, M. Jufer, and F.-L. Perret. 1991. Transport Interrégional à Grande Vitesse—Le Projet SWISSMETRO. Lausanne: Ecole Polytechnic Fédérale de Lausanne.

Nye, D. 1990. Electrifying America. Cambridge, Mass.: MIT Press.

Pennsylvania Power and Light. 1940. Origin and Development of the Company. Vol. 1. Corporate History in nine volumes. Allentown, Penn.: Pennsylvania Power and Light.

Poe, E. A. 1976. The Science Fiction of Edgar Allen Poe. New York: Penguin.

Sack, J. B. 1995. The Darwinian Theory of Distribution. Global Electricity Strategy Research Memorandum Series. December 12. New York: Morgan Stanley.

Sack, J. B. 1996. An Interrelated World. Global Electricity Strategy Research Memorandum Series. February 29. New York: Morgan Stanley.

Schilling, H. D., and R. Hildebrandt. 1977. Primarenergie-Elektrische Energie, Die Entwicklung des Verbrauchs an Primarenergietragern und an Elektrischer Energie in der Welt, in den USA und in Deutschland seit 1860 bzw. 1925. Essen, Germany: Vertrag Gluckäuf.

Schivelbusch, W. 1988. Disenchanted Night: The Industrialization of Light in the Nineteenth Century. Berkeley, Calif.: University of California Press.

Shurr, S. H., C. C. Burwell, W. D. Devine, Jr., and S. Sonenblum. 1990. Electricity in the American Economy: Agent of Technological Progress. Westport, Conn.: Greenwood Press.

Starr, C. 1995. A personal history: Technology to energy strategy. Annual Review of Energy and the Environment 29:31–44.

Thirring, H. 1958. Energy for Man. Bloomington, Ind.: Indiana University Press.

US Bureau of the Census. 1978a. Historical Statistics of the United States. Washington, D.C.: US Bureau of the Census.

US Bureau of the Census. 1978b. Statistical Abstract of the United States. Washington, D.C.: US Bureau of the Census.

US Bureau of the Census. 1981. Statistical Abstract of the United States. Washington, D.C.: US Bureau of the Census.

US Bureau of the Census. 1984. Statistical Abstract of the United States. Washington, D.C.: US Bureau of the Census.

US Bureau of the Census. 1986. Statistical Abstract of the United States. Washington, D.C.: US Bureau of the Census.

US Bureau of the Census. 1989. Statistical Abstract of the United States. Washington, D.C.: US Bureau of the Census.

US Bureau of the Census. 1991. Statistical Abstract of the United States. Washington, D.C.: US Bureau of the Census.

US Bureau of the Census. 1992. Statistical Abstract of the United States. Washington, D.C.: US Bureau of the Census.

US Bureau of the Census. 1994. Statistical Abstract of the United States. Washington, D.C.: US Bureau of the Census.

Technological Trajectories and the Human Environment. 1997.
Pp. 135–156. Washington, DC: National Academy Press.

Materialization and Dematerialization: Measures and Trends

IDDO K. WERNICK, ROBERT HERMAN,
SHEKHAR GOVIND, AND JESSE H. AUSUBEL

INTRODUCTION

"Revenge theory" postulates that the world we have created eventually gets even with us, twisting our cleverness against us (Tenner, 1996). Helmets and other protective gear have made American football more dangerous than its bare predecessor, rugby. Widened roads invite more vehicles, which mitigate gains in average traffic speed and flow. In short, human societies face unintended and often ironic consequences of their own mechanical, chemical, medical, social, and financial ingenuity.

In 1988 Robert Herman, Siamak Ardekani, and Jesse Ausubel began to explore the question of whether the "dematerialization" of human societies is under way (Herman et al., 1989). At that time, dematerialization was defined primarily as the decline over time in the weight of materials used in industrial end products or in the "embedded energy" of the products. More broadly, dematerialization refers to the absolute or relative reduction in the quantity of materials required to serve economic functions.

Dematerialization matters enormously for the human environment. Lower materials intensity of the economy could reduce the amount of garbage produced, limit human exposures to hazardous materials, and conserve landscapes. From time to time, fears arise that humanity will imminently exhaust both its material and energy resources. Historically, such fears have proven exaggerated for the so-called nonrenewable resources such as metals and oil. Yet if the human economy were to carelessly metabolize large amounts of Earth's carbon or cadmium, the health and environmental consequences could be dire. Meanwhile, the so-called renewable resources, such as tropical woods, are proving difficult to renew when demand is high. Thus, a general trajectory of dematerialization would certainly favor sustaining the human economy over the long term.

Is dematerialization occurring? Certain products, such as personal computers and beverage cans, have become smaller and lighter over the years. However, revenge effects may still countervail. A vexing case is that total paper consumption has soared despite claims that the electronic information revolution would create a paperless office. Americans now use about a kilogram of paper per day on average, twice the amount used in 1950.

In this essay we report further analyses of materialization and dematerialization, mostly for the United States during this century, and lay the basis for a systemwide assessment. We segment our analysis to consider measurements: (1) at the stage of resource extraction and the use of primary materials, such as minerals, metals, and wood; (2) in industry and industrial products; (3) at the level of the consumer and consumer behavior; and (4) in terms of the waste generated. At each stage one can ask whether dematerialization is taking place, what drives it, and what are its future trajectories and their consequences. Our studies consider materials in absolute terms, per unit of economic activity (measured by means of gross national product, GNP, or its slight "domestic" variant, GDP), and per capita. We assess changes in both volume and weight.

Materials consumption is analytically less tractable than energy use. It cannot be satisfactorily reduced to single elementary indicators such as kilowatt-hours or British thermal units. To illustrate this point, a pound of gold cannot be simply compared with a pound of lead, to the frustration of the alchemists. And neither one can be easily compared with a pound of plutonium. Materials possess unique *properties,* and those properties provide value, define use, and have environmental consequences. To capture these and other interactions, we must consider an ensemble of measures under the rubric of dematerialization.

The pattern of materialization and dematerialization, and the database from which it is drawn, helps frame the new field of industrial ecology (Frosch, 1992). Industrial ecology is the study of the totality of the relationships between different industrial activities, their products, and the environment. It is intended to identify ways to optimize the network of all industrial processes as they interact and live off each other, in the sense of a direct use of each other's material and energy wastes and products as well as economic synergism. The macroscopic picture of materialization can help raise key research questions and set priorities among the numerous studies of materials flows and networks that might be undertaken. It puts these in a dynamic context of both technical and market change.

DEMATERIALIZATION AND PRIMARY MATERIALS

In analyzing primary materials, it is helpful to begin with a profile of the total "basket of stuff" that a human society consumes. For this purpose, let us consider first "demandite," an imaginary, composite material representative of the nonrenewable resources we use. Demandite reveals our elemental preferences. All the materials that make up demandite are quantified in terms of the total moles of each element (or selected compounds) they contain; demandite is characterized by the fraction of moles for each element or compound divided by the total mole number.

First proposed by Goeller and Weinberg (1976), demandite includes both the energy materials (the hydrocarbon fossil fuels: coal, oil, and gas) and other materials, such as iron, copper, sulfur, and phosphorus, that are mined and used in the production of goods. Demandite omits some (crushed) stone that is used to build roads and other structures; the amount mobilized is quite large (about 1 billion metric tons in the United States in 1990) but this stone resource is practically infinite, and its elemental composition is more or less the average of the Earth's crust. Following Goeller and Weinberg, we recalculated the percentage of moles of the composite materials in demandite for the United States for 1968 and estimated these percentages for 1990 (Table 1).[1]

The hydrocarbon compounds dominate demandite. They swelled from about 83 percent of US demandite in 1968 to over 86 percent in 1990. In fact, total US consumption of hydrocarbons in 1990 was over 1.9 billion metric tons (t) or about 7.8 t per capita (20 kg per capita per day). The extraction and use of

TABLE 1   Demandite (United States)

Elements and Compounds	Percentage of Total Moles	
	1968	1990
Hydrocarbons	83.20	86.77
Silicon dioxide	12.33	9.35
Iron	1.30	0.64
Oxygen	0.61	0.76
Sodium	0.57	0.44
Chlorine	0.57	0.44
Nitrogen	0.47	0.67
Phosphorus	0.29	0.35
Sulfur	0.21	0.25
Calcium carbonate	0.15	0.10
Aluminum	0.15	0.13
Potassium	0.09	0.07
Copper	0.04	0.02
Zinc	0.02	0.01
Lead	0.005	0.004
Magnesium	0.004	0.003

SOURCES: Chemical Manufacturers Association (1991), US Bureau of the Census (1975, 1992), and US Bureau of Mines (1976).

hydrocarbons pose problems such as global warming and oil spills, as well as health threats from urban, vehicular, and groundwater pollution.

The carbon, not the hydrogen, of course, is the "bad" element in the environmental story. The "decarbonization" of the economy is thus clearly of paramount environmental importance. As discussed by Nakićenović (see Nakićenović, this volume), relative to GNP and energy production, decarbonization is occurring steadily. Yet absolute carbon consumption by weight in the United States grew at a compound rate of 1.8 percent per year between 1950 and 1993.

Excluding energy materials, US material flows, including crushed stone and all other physical materials as well as renewables (with the exception of food), amounted to about 2.5 billion metric tons in 1990 or about 10 t per person (28 kg per capita per day) (Rogich et al., 1993). Construction materials dominated with 70 percent of total apparent US consumption (Table 2). Materials in this category may be associated with local environmental issues as excavations and structures transform the landscape, for better or worse. A striking fact is that 30 percent of the industrial minerals consumed dissipated into the environment and thus were rendered practically unrecoverable.

TABLE 2   Nonfuel Materials Flows in the United States, 1990

Material Group	Apparent Consumption (10^6 metric tons)	Recycled Quantity (%)	Dissipative Use (%)	Postconsumer Waste (%)	Processing Waste (%)
Construction Minerals	1,746	8	2	8	4
Industrial Minerals	330	8	30	8	2
Metals	112	55	0.2	13	5
Nonrenewable Organics (e.g., plastics)	112	2	---	19	---
Renewable Organics (e.g., forest products)	231	8	0.4	34	---
Animal Products (e.g., hides)	2.2	1	76	2	1

NOTE: The recycled as well as other materials flows shown are those reported for the same year as the apparent consumption and do not account for the time lag associated with the actual reprocessing or other endpoints of those materials.
SOURCE: Rogich et al. (1993).

TABLE 3 Material Densities in Metric Tons/Cubic Meter

Air @ 20°C 1 ATM	0.00121
Spruce (avg.)	0.43
Oak (avg.)	0.68
Gasoline	0.68
Commodity Polymers	0.9–1.25
Water	**1.00**
Whole Blood	1.06
Bone	1.7–2.0
Concrete	2.3
Granite	2.7
Glass, common	2.4–2.8
Earth's crust (avg.)	2.8
Aluminum	2.7
Steel	7.8
Copper	8.9
Silver	10.5
Lead	11.3
Mercury	13.6
Gold	19.3
Platinum	21.4
Osmium	22.5

SOURCES: Giancoli (1988) and *Handbook of Chemistry and Physics* (1962).

Fluctuations but no trend in absolute consumption by weight of physical materials are evident for the United States for the past twenty-five years (Rogich et al., 1993). However, an assessment of consumption per unit of economic activity shows a dematerialization in physical materials of about one-third since 1970. The oil shocks of 1973 and 1979 appear to have ratcheted the ratio down.

A complete current accounting of materials consumption in America, including renewables, nonrenewables, and both energy and nonenergy materials, reveals a total of more than 50 kg per capita per day (Wernick and Ausubel, 1995). (We have excluded the consumption of water and air.) This equals about 20 t/yr. If we assume a constant consumption over eighty years, an individual American's lifetime consumption would be 1,600 t. If the average material has the density of water (Table 3), the volume of material consumed in one person's lifetime would be equivalent to a cube measuring about 13 meters on a side.

Disaggregation reveals that the intensity of use of diverse materials has

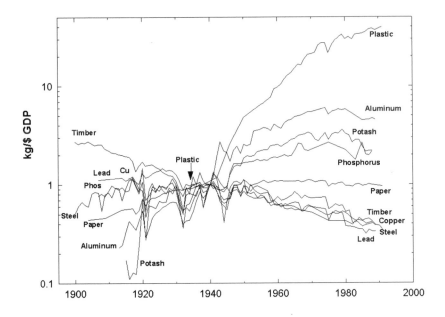

FIGURE 1 Intensity of use of materials in the United States. NOTE: Consumption data are divided by GDP in constant 1987 dollars. (For example, in 1907 the United States consumed 405,312 metric tons of lead and GDP was $371.28 billion, giving a ratio of about 1.09 metric tons of lead per million dollars GDP; in 1990 1,220,000 metric tons of lead were consumed and GDP was $4,152 billion, giving a ratio of about 0.29 metric tons per million dollars GDP.) For plastics we use production data only. DATA SOURCES: *Modern Plastics* (1960); US production of plastics resin, personal communication with Joel Broyhill, Statistics Department, Society of the Plastics Industry, Washington, D.C., August 20, 1993; US Bureau of the Census (1975, 1991, 1992, 1993, 1994, 1995); and US Bureau of Mines (various years).

changed dramatically over the twentieth century (Figure 1) (see also Ross et al., 1985). In terms of the weight of material used, normalized by GDP, timber sloped steadily down from its top position in 1900. Steel, copper, and lead also slid from their earlier heights. Plastics and aluminum followed upward trajectories, as did phosphates and potash, key ingredients in agricultural fertilizers.

Wood remained the preeminent material in the United States into the 1930s. It was used for fuel and as the structural material to build homes, workshops, vehicles, and bridges. It provided both ties and rolling stock for railroads as well as lamp, electric, and telephone poles for the utility infrastructures. Wood was gradually replaced by other materials and made more durable by creosote and other preservatives. Since 1930, annual US per capita consumption of commercial lumber has remained stable at about 200 board feet, down from about 500 board feet per capita at the turn of the century. An 80-foot spruce measuring 15 inches in diameter at breast height typically yields 200 board feet and is about sixty years old. The wood pile in earlier times was much larger, of course, because of the noncommercial use of wood for fuel.

In 1900, less than 2 percent of the timber cut in the United States produced pulp and paper. Today that fraction is over 25 percent. Absolute paper consumption has climbed steeply (Figure 2), while consumption per capita has risen more slowly. Amidst the electronic revolution, paper remains a preferred carrier of information. New technologies for information storage supplement the range and augment the amount of information stored, rather than reduce the use of paper. But, as Figure 2 also shows, paper consumption per unit of GNP has stayed essentially flat since 1930. During World War II, America briefly and drastically reduced its paper use relative to GNP or, rather, increased GNP without increasing its paper use.

Newly exploited materials now complement traditional ones, fortifying and enhancing their properties like vitamins. Materials such as gallium, the platinum-based group, vanadium, and beryllium have come into use in electronics and in the production of steel alloys and other "designer materials." The absolute amounts are small (10^{-4}–10^{-2} kg per capita per year) and relatively steady (US Bureau of Mines, various years). The small amounts of these new materials understate their importance both economically and environmentally. Extensive processing before final use may involve large ore bodies and mine wastes. Near

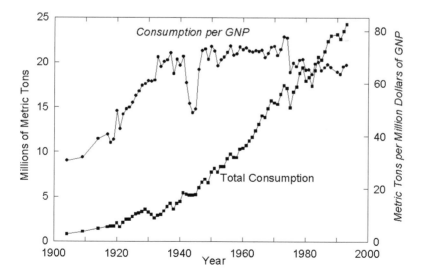

FIGURE 2 Absolute paper consumption and paper consumption per unit of GNP in constant 1982 dollars. DATA SOURCES: W. E. Franklin and Associates (1990) and US Bureau of the Census (1975, 1991).

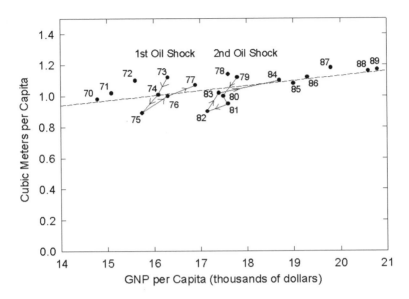

FIGURE 3 US per capita volumetric material consumption for wood/plywood products, paper/paper products, plastic, and metals plotted against GNP per capita in constant 1990 dollars. NOTE: Volume is calculated from average material densities. The solid lines show the short-lived effects of the oil shocks. The dashed line is a linear best fit to the historical data. SOURCE: After Rogich et al. (1993).

optimal use is then made of the elemental physical and chemical properties of the refined materials. Precisely because of their potency, these are distributed in small amounts that are sometimes difficult to recover, thereby frustrating efforts to recycle other materials "contaminated" by them.

Although the weight of materials employed may be stable or declining, the volume of materials is gradually increasing in the United States. Average density values (Table 3) can be combined with consumption data for individual materials to estimate the total volume of materials consumed. As shown in Figure 3, since 1970 volume per capita of the combination of paper, wood, metals, and plastics has increased along with economic growth, according to a linear best fit of the data and notwithstanding the oil price shocks of 1973 and 1979. Within five years of each shock, the system had resumed its long-term volumetric expansion. Individual items in the American economy may be getting lighter, but the economy as a whole is physically expanding.

Plastics, as a result of their increased consumption, account for much of the growth in volume; they are the preferred lower density materials. Polymer plastics are a "manufacturer friendly" material because they can be shaped into complex geometric forms with relative ease, are chemically inert, and can be created with a wide range of properties. Plastics have occupied market niches from car bumpers to soft-drink containers to furniture and plumbing parts.

Although commercial plastics were introduced early in the century (Bakelite, in 1909), they did not seriously enter the economy until 1940. About 500 billion kilograms of plastic have been produced so far in the United States, or, in terms of the current US population, about 2,000 kg per capita (*Modern Plastics*, 1960). Extrapolating the historical production, we estimate that the cumulative amount of plastic resin produced in the United States will roughly double by 2030. The primary feedstocks for plastic are oil, the dominant hydrocarbon, and, more recently, natural gas. One might say that plastics have been a by-product of the automobile. As long as cars run on oil, new plastic resin will be cheaply available. However, the decarbonization of the energy system and the growth of the plastic endowment will encourage much greater recycling of plastics over the next three or four decades, countering the mounting problem of plastic waste disposal. The high level of customization of plastics complicates matters, as this diversity makes effective separation and reuse more difficult.

DEMATERIALIZATION IN INDUSTRY AND INDUSTRIAL PRODUCTS

We can readily assess two relevant aspects of the materials used in industry: the dematerialization of individual end products and the use of recycled materials in production. More complete "life-cycle analyses" of products must embrace such partial examinations and extend them. For example, knowing the complete material and energy demand of a typical milk container throughout this century would be revealing. At present, we are unaware of life-cycle analyses repeated over time that provide an indication of trends.

Several individual end products manifest dematerialization. Containers, for example, have generally become lighter. At mid-century, beverage containers were predominately made of steel or glass (US Bureau of Mines, 1990). In 1953 the first steel soft-drink can was marketed. The public accepted it, resulting in the erosion of the market share of the heavier glass containers. Cans of aluminum, a material one-third the density of steel, entered the scene a decade later and grew from a 2 percent market share in 1964 to almost 90 percent of the soft-drink market and about 97 percent of the beer market by 1986. The aluminum can was itself lightened by 25 percent between 1973 and 1992 (Garino, 1993). In 1976, polyethylene terephthalate (PET) resins began to occupy a significant portion of the market, especially for large containers, where glass had previously dominated.

Cars have also become lighter on average, although the recent sales growth of light trucks and sport vehicles counters this trend. The car is an interesting object for study because it represents a full market basket of the products of an industrialized economy, including metals, plastics, electronic materials, rubber,

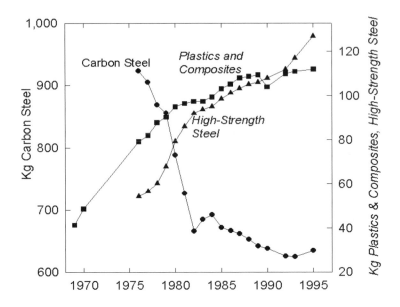

FIGURE 4. Changing weight of selected materials in the average US automobile. DATA SOURCE: *Wards Automotive Yearbook* (1969–1995).

and glass. Cars have become more materially complex as well, an important factor in the difficulty of disassembly and reuse. Only recently have legislative and engineering efforts, particularly in Germany, been directed toward the design and eventual production of car components that can be replaced and recycled with minimal effort. In the early 1970s, the amount of carbon steel in the average US car began to decline and then fell sharply by about 300 kg or 35 percent (Figure 4). A combined increase of about 100 kg in plastics and composites and in high-strength steel helped "downmass" the automobile while maintaining its structural integrity, with the new materials substituting for the old in a ratio of about one to three.

Although aircraft use a tiny fraction of all materials, the aerospace sector, where performance demands are exceptionally stringent, has foreshadowed trends that later appear in the rest of the economy. The drive to downmass aircraft while improving their performance has placed the aerospace industry at the forefront of materials research. Each kilo safely shaved in design saves both fuel and money. The aerospace industry provides the strongest market for composite materials of carbonaceous fibers mixed with aluminum and titanium, which have excellent strength-to-weight ratios. Materials with greater thermal resistance also yield higher performance. An increase of 80°C in the operating temperature of a jet engine can yield a 20 percent increase in engine thrust (US Bureau of Mines, 1990). Engines once formed from pure nickel now include nickel and cobalt-based superalloys, aluminum-lithium alloys, and ceramics, all having superior thermal properties. Specialized coatings and paints used on aircraft now provide significant environmental handling challenges. The mounting trend here is toward complexity, not only in the final product, but in the processing stage as well as ultimately in disposal.

The closure of materials loops through the reuse of materials complements the downsizing route to dematerialization. While smaller and lighter products can reduce the amount of materials required by future generations to operate the economy, reuse and recycling can also minimize fresh inputs and waste outputs (Wernick, 1994). At present, secondary inputs to production have difficulty competing with virgin materials in many markets.

Successful secondary materials recovery relies on two basic factors: ease of isolation of the desired materials and consumer demand for reprocessed materials. The difficulty of isolation explains why only 7 percent of cadmium-loaded waste was recycled from hazardous waste streams in 1986 (Allen and Behmanesh, 1994) and an even smaller fraction of arsenic and thallium. The ease of isolation explains why lead now enjoys a recovery rate exceeding 70 percent of its demand. Lead is used mainly for automobile batteries, which are readily separated from the general waste stream. Dissipative uses of lead (e.g., paint and gasoline) have been substantially curtailed over the last few decades. The supply of wastepaper, also easily separated, has proved responsive to market demand, with a growing number of paper mills now accepting this source of fiber.

Steel is another material that is readily separated. The secondary supply neatly meets the demand created in large part by the technology of electric arc steel production, which relies primarily on scrap inputs. Electric arc steel production has risen steadily throughout the century and now constitutes about 40 percent of all steel production. Trace contaminants in scrap piles may confound future gains in this method of steel production, however. Contaminants, such as zinc, can be problematic even at a level of tens of parts per million and can result in substandard finished steel.

Demand for secondary materials, like all demand, is to a large degree a function of price. Precious metals such as gold and silver are commonly recovered from circuit boards. However, even the recovery of precious metals has limits. In the late 1980s, platinum prices needed to exceed $500 an ounce to make platinum recovery from catalytic converters economical (Frosch and Gallopoulos, 1989). Energy prices can spark secondary materials recovery trends in opposite directions. Recycled aluminum requires only 5 to 10 percent of the energy necessary for primary production, and this difference has spurred secondary recovery when energy is costly. At the same time, cheap and plentiful energy is often the obvious technical requirement for economical reversal of the dissipation of many materials, even when diluted to concentrations found in sea water.

The fraction of the total production of given materials supplied by secondary materials recovery broadly indicates the extent to which the economy functions

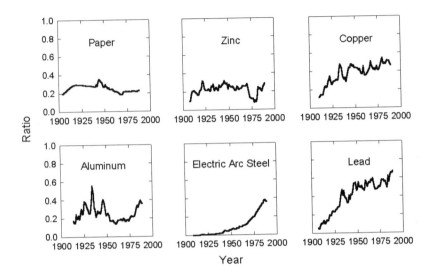

FIGURE 5. Ratio of production for secondary to primary sources of paper, zinc, copper, aluminum, steel, and lead in the United States. DATA SOURCES: US Bureau of the Census (1975, 1991).

with closed materials cycles (Figure 5). This measure shows little change for paper and zinc since early in the century. Copper rose rapidly in the early part of the century but has not been able to sustain an upward trend since 1940. Aluminum jumped in the late 1970s but may have plateaued again. Steel has climbed and points toward 50 percent. As noted, lead has surpassed 70 percent. World War II elicited peaks of materials reuse in the United States for several materials. These levels of reliance on recovered materials, often about 40 percent, may represent the current practical upper limit for many materials. The reasons appear to be not so much physical as economic. For example, suppliers of virgin materials can adjust their prices to undercut recycled supplies. Economic and population growth, of course, tend to draw new materials into the system.

DEMATERIALIZATION AND CONSUMERS

The number of consumers and their individual and collective behaviors drive materialization. An obvious fact is that there are more and more consumers. During the twentieth century the population of the United States has more than tripled, from about 80 million to more than 250 million. The absolute number is only part of the story. Life-styles also shape demand. Today, only a small fraction of consumption in wealthy nations (or communities) is actually for basic survival; most is for pleasure and to express one's standing in society.

Although fast modes of travel have enabled settlements to spread, Americans on average appear to be increasing the density of their built environment over time. An analysis of data from many diverse neighborhoods in Austin, Texas, indicates that the average size of a residential plot decreased between 1945 and 1985 (Figure 6). Meanwhile, the average floor area of an Austin residence increased by 50 percent or more. The fraction of a plot covered with a structure also increased by about 50 percent in both the mean and the aggregate.

A steady increase in land value over the years could contribute to the decline in plot area in some zones. "Market forces" may adjust the size of the plot area to reduce the impact of higher costs of land. In any new development, most of the land parcels are not portioned off by prospective buyers but rather by the developers. Hence, apart from making a binary decision whether or not to purchase, the buyer has limited influence on the subdivision planning and design process. The process of fixing the size of plots may be driven primarily by the economics of the situation as perceived by the developer.

Our hankering for a domicile in idyllic settings was what drove us to suburbia. Contrary to conventional belief, once we get there, we do not seem to care about how small the plot area is. Notwithstanding professed tastes for open space, we seem to build, enclose, and accrete steadily. We also seem to display a prefer-

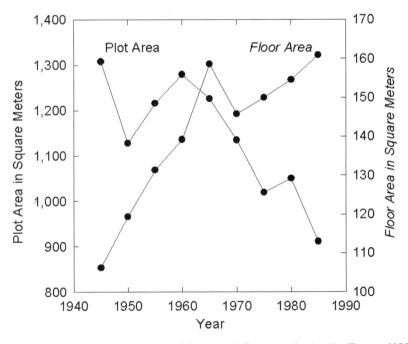

FIGURE 6 Size of average residential plot and floor area in Austin, Texas. NOTE: Data records for all single-family dwellings built during the period 1945–1990 in twenty-five of the twenty-nine zip codes within the Austin metropolitan area were used. DATA SOURCE: Austin Multiple Listing Service, Inc.

ence for larger and larger floor areas. This would also suggest that we have enough money to spend on larger houses or the material goods to fill up all the new (or "extra") floor area, but we are not perceived as having the desire to spend on larger plot areas.

Today's enlarged homes house fewer people. The number of residents per housing unit has declined monotonically in the United States from five in 1890 to fewer than three today (Figure 7). Interestingly, when the increase in floor area over time is viewed in conjunction with the decrease in the average number of residents per occupied housing unit, one could conclude that the *floor area available per person almost doubled in forty-five years.* This hypothesis should be treated with caution because it combines national aggregate data with data only for the city of Austin.

The trend toward larger floor areas housing fewer people implies that our consumption of building materials on a per capita basis is increasing, as is our requirement for energy services such as heating and cooling (see Schipper, this volume).

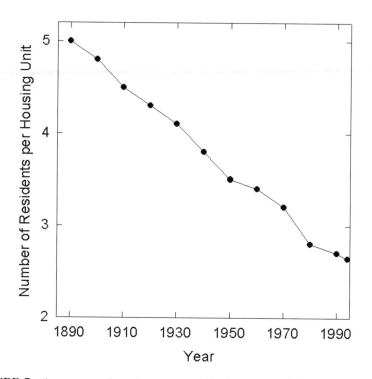

FIGURE 7 Average number of persons residing in an occupied housing unit in the United States (farm and nonfarm). DATA SOURCES: US Bureau of the Census (1975, 1995).

National data on the weight of household moves corroborates the Austin data of an increase in floor area. Data on the weight of household goods transferred in intercity moves implies that Americans on average possess more stuff over time. The average load increased by almost 20 percent to 3,050 kg (about a sixty-day stock of accumulated materials at the American rate of 50 kg/day) from 1977 to 1991 (personal communication from George Bennett, Statistics Department, Household Goods Carriers Bureau, Alexandria, Va., October 14, 1993). Intercity freight transport, measured in ton-miles, also aggregates a great deal of material. This measure indicates that as time goes by, Americans either ship more tons, ship each ton farther, or both. In 1990 Americans on average shipped freight 11,000 ton-miles; this compares with about 7,200 ton-miles in 1950. An increase in the weight, and therefore very likely the bulk, of possessions moved implies a need for more household space. Conversely, an increase in household floor space would always be filled by our insatiable appetite for possessions.

We also ship more information. The number of pieces of mail per capita in the United States has more than tripled since 1940 and stands today at more than 650 pieces per capita annually. Conservative estimates of the growth rate of electronic mail transmission and use in the United States is a phenomenal 10 percent per month. The information economy does not appear to substitute for the materials economy but may rather be required to manage its growth.

Whatever the reasons, from the vantage point of building materials, structures, and household amenities per capita, wealth appears to be a materializer. The shift from larger to smaller families materializes. Each housing unit is built of materials and in turn is filled with more objects to be used by fewer people.

The "individuation" of products also materializes. The dominant American social model is not simply to own one's home but to have access to products and services customized for more and more niches. Packages of "portion-controlled" prepared food and, more generally, product proliferation exemplify individuation. From 1980 to 1993 the number of new products introduced in supermarkets grew at an average compound rate of 14 percent per year. More than seventeen thousand new items appeared on store shelves in 1993 (personal communication from Lynn Dornblaser, Statistics Department, New Product News, Chicago, June 8, 1994).

Although many products saturate the consumer markets temporarily, they often rematerialize at a higher level. One example is the telephone. During the 1930s, telephones saturated in the United States at a rate of two for every ten persons (US Bureau of the Census, 1975). Starting about 1940, phones found new markets—the result of prosperity, new functions and fashions, and better performance. After 1970 it became difficult to keep track of the number of phones in operation in the United States. It is not surprising that the Bell System crumbled when it contained more than one hundred million objects at the level of the end user. Numerous phones now ring for every American. A continuation of this trend will bring the United States to many hundreds of millions of devices by

2020. On average each new generation of devices is smaller and lighter than its antecedents and performs more functions, such as fax transmission, voice mail, and wireless mobile telecommunications. An interesting question is whether the total mass of the telecommunications system, including cables and other equipment and facilities, has changed much since its initial formation earlier in the century. Revenge theory suggests that the overall system growth may offset the efficiency gains in its components.

DEMATERIALIZATION AND WASTES

Poor data, unreliable and inconsistent categorization, and infrequent surveys impede the establishment of waste trends (Rathje and Murphy, 1992; US Congress, Office of Technology Assessment, 1992). Serious interest in comprehensive rubbish data is quite recent. Moreover, some material by-products, which might otherwise flow into further productive use, become waste due to rather arbitrary labeling laws and regulations (see Frosch, this volume). One year for which comprehensive data for the United States are available is 1985 (Figure 8). Industrial wastes dominate, but 90 percent of industrial (including manufacturing) wastes can be water, so a comparison between classes may be misleading.

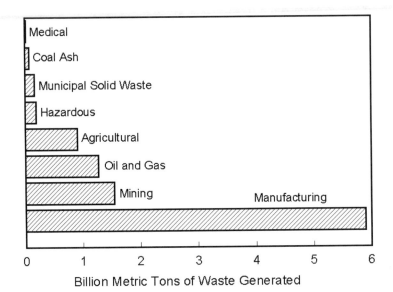

FIGURE 8   Major waste types by weight in the United States, 1985. NOTE: A large fraction of the total weight in the industrial categories is water. Dry weight of industrial wastes can be as low as 10 percent of the total. DATA SOURCE: US Congress, Office of Technology Assessment (1992).

When we simply add all of the wastes, the total in 1985 was about 10 billion metric tons, or about 115 kg per capita per day. Because a large and unknown fraction of this amount is water, it cannot be compared with our 50 kg per capita per day estimate of material use. Further research might lead to such a comparison and rough total guesses of net long-lived materialization. Instead, here we cautiously provide some comparisons of waste categories over time.

Sewage sludge almost doubled between 1972 and 1992 in the United States to 5.4 million dry metric tons, about 21 kg per capita annually (Jin, 1993). The increase does not indicate a change in human metabolism but rather in population growth and increased treatment of waste. The main source of *ash* is coal power plants. A rule of thumb is that about 10 percent of all coal burned remains as bottom ash, boiler slag, or fly ash captured by air pollution devices. As coal consumption declines, so does ash production, though flue-gas desulfurization may increase sludge while decreasing air pollution. Inclusion of coal and wood ash in historical analyses could substantially modify the picture of waste trends; it would flatten the recent rise by increasing the amount of waste generated in earlier periods. In 1990, the amount of coal ash produced equaled about 350 kg per American. *Hazardous wastes* include several hard-to-define subcategories, and no long-term figures are available (Allen and Jain, 1992). For such wastes environmental effects may sometimes be measured in micrograms rather than megatons. According to the US Environmental Protection Agency (EPA), in 1985 total hazardous waste generation was 271 million metric tons, while in 1987 it was 238 million, or roughly 1 t per capita. A related category is *toxic chemicals* and compounds released to air, water, and land by industrial facilities. Allowing for considerable uncertainty, EPA reports indicate that the total amount declined from about 2.2 million metric tons (4.8 billion pounds) in 1988 to 1.4 million metric tons (3.2 billion pounds) in 1992, or about 5.6 kg per American (INFORM, 1995).

In absolute terms, *municipal solid wastes* (MSW) in the United States rose from 80 million metric tons in 1960 to 188 in 1993, or about 725 kg per capita (US Bureau of the Census, 1995; US Environmental Protection Agency, 1992). About one-third of MSW by weight consists of packaging products. While MSW increased by about 1.5 percent per capita annually in the United States between 1960 and 1993, the amount of American trash generated per unit of GNP decreased, or dematerialized, on average by about 0.3 percent per year despite a recent upturn (Figure 9). As shown in Figure 10, reported waste generation varies markedly by country. Even among advanced industrialized nations, reported waste varies by up to a factor of three, which may partly reflect differences in categorization. The German data show a tight lid on trash, and in fact Germany seeks to reduce its packaging waste by 80 percent (Fishbein, 1994). Although it is hard to make global generalizations, the United States by all measures stands apart in its high level of trash generation.

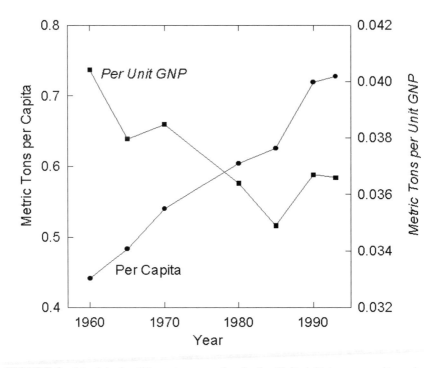

FIGURE 9   Municipal solid waste generation in the United States per capita and per unit of GNP in constant dollars. DATA SOURCES: US Bureau of the Census (1975, 1995) and US Environmental Protection Agency (1992).

QUESTIONS AND CONCLUSIONS

Is it possible to bring together existing evidence in a general theory of materialization and dematerialization? Bernardini and Galli (1993) have proposed a two-part theory that merits attempts at validation. The first part of their theory is that new materials substitute for old in subsequent periods of time (Fisher and Pry, 1971), and each new material shows improved physical properties per unit quantity, thus leading to a lower intensity of use. The second part of the theory applies to the development of nations or regions. The concept is that countries complete phases of their development sequentially at roughly the same value of per capita GDP, but the intensity of use of a given material declines depending on when each country completes its development, as the late-arriving economies take advantage of learning curves.

The Bernardini and Galli theory is hopeful for dematerialization. It implies that continued research and development of materials will bring about substantial gains and that global development will not dumbly imitate the behavior of the early developing nations. Thus, for example, China and India will never repro-

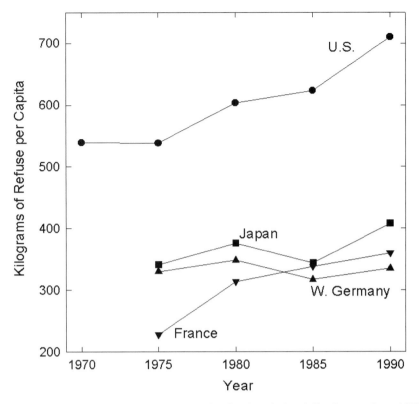

FIGURE 10 Refuse generation per capita for four industrialized countries. NOTE: Waste categorization and measurement vary considerably from country to country. DATA SOURCES: OECD (1991, 1994).

duce the pattern of per capita materials flows of the United States, even if their GDPs grow dramatically.

While appreciating this general logic for dematerialization, what does the actual evidence show so far for the United States? Our survey suggests the following:

(1) With regard to primary materials, summary ratios of the weight of materials used to economic product appear to be decreasing due to materials substitution, efficiencies, and other economic factors. The tendency is to use more scientifically selected and often artificially structured materials (Sousa, 1992). These may be lighter, though not necessarily smaller. The value added clearly rises with the choice of material, but so may aggregate use.

(2) With regard to industry, encouraging examples of more efficient materials use exist in many sectors, functions, and products. Firms search for opportu-

nities to economize on materials, just as they seek to economize on energy, labor, land, and other factors of production. However, the taste for complexity, which often meshes with higher performance, may intensify other environmental problems, even as the bulk issues lessen.

(3) As consumers, we profess one thing (that less is more) and often do another (buy, accrete, and expand). We see no significant signs of net dematerialization at the level of the consumer or saturation of individual material wants.

(4) With regard to wastes, recent, though spotty, data suggest that the onset of waste reduction and the rapidity with which some gains have been realized as well as the use of international comparisons indicate that very substantial further reductions can take place.

An overall assessment clearly requires an ensemble of measures under the rubric of dematerialization. System boundaries matter a great deal, whether they be nations, regions, economic sectors, firms, households, or products. In general, viewing the environmental impact of a product in isolation from the total system is simplistic. For example, examined in isolation, a computer could be deemed environmentally unfriendly because the production of the printed circuit boards, logic and memory chips, and display screens requires a large quantity of hazardous chemicals and solvents and heavy metals. This conclusion could also be based on its propensity for the consumption of paper and energy. However, the operation of the same computer in an industrial setting could increase efficiencies in a manufacturing process and reduce the consumption of energy and raw materials and the generation of waste. What a cursory scrutiny might identify as a local maximum could be a global minimum in terms of adverse environmental impacts. Thus, we must measure products, product life cycles, sectors, and the total materials economy at several stages, denominated in various ways.

A logical next step in research is to develop a self-consistent scenario for a significantly dematerialized economy and to explore the changes in technology and behavior needed to achieve it. Such an exercise should include careful examination of hazards as well as benefits to the environment associated with a qualitatively and quantitatively new materials economy.

Clearly, intensive research in materials science and engineering, sensitive to environmental properties, is a key to dematerialization. Humanity has passed from the ages of stone, bronze, and iron to an age in which we deliberately employ all ninety-four naturally occurring elements of the periodic table. We must learn to use the elements even better, in a way compatible with our long-term well-being.

In conclusion, we return to the observation that substantial progress has been made over the past century in decoupling economic growth and well-being from increasing primary energy use through increased efficiency. Decoupling materials and affluence will be difficult—much harder than decoupling carbon and prosperity. Objects still confer status, and they take their revenge.

ACKNOWLEDGMENT

We gratefully acknowledge assistance from John D. C. Little, Perrin Meyer, and Umer Yousafzai. We also thank the Austin Multiple Listing Service, Inc. and Borris Systems for providing access to their database of Austin residences.

NOTE

1. Our recalculation of demandite for 1968 differs from the results obtained by Goeller and Weinberg (1976). We believe that the discrepancy is due to our using different data inputs for calculating the silicate and calcium carbonate contributions to demandite. Our result of 12.33 percent for SiO_2 is slightly larger than their 11.15 percent figure, and our result of 0.15 percent form $CaCO_3$ is considerably smaller than their figure of 4.53 percent. We note that this disparity in results does not alter the conclusions drawn from the table, as it hardly affects the dominant position of the hydrocarbons, and environmental effects of these two minerals are essentially equivalent and benign.

REFERENCES

Allen, D. T., and N. Behmanesh. 1994. Wastes as raw materials. Pp. 69–89 in The Greening of Industrial Ecosystems, B. R. Allenby and D. J. Richards, eds. Washington, D.C.: National Academy Press.

Allen, D. T., and R. K. Jain, eds. 1992. Special issue on national hazardous-waste databases. Hazardous Waste & Hazardous Materials 9(1):1–111.

Bernardini, O., and R. Galli. 1993. Dematerialization: Long-term trends in the intensity of use of materials and energy. Futures (May):431–448.

Chemical Manufacturers Association. 1991. United States Chemical Industry. Washington, D.C.: Chemical Manufacturers Association.

Fishbein, B. K. 1994. Germany, Garbage, and the Green Dot. New York: INFORM.

Fisher, J. C., and R. H. Pry. 1971. A simple model of technological change. Technological Forecasting and Social Change 3:75–88.

Frosch, R. A. 1992. Industrial ecology: A philosophical introduction. Proceedings of the National Academy of Sciences of the United States of America 89(3):800–803.

Frosch, R. A., and N. E. Gallopoulos. 1989. Strategies for manufacturing. Scientific American 260:144–152.

Garino, R. J. 1993. Can recycling, back on track. Scrap Recycling and Processing (May/June).

Giancoli, D. C. 1988. Physics for Scientists and Engineers. Englewood Cliffs, N.J.: Prentice-Hall.

Goeller, H. E., and A. M. Weinberg. 1976. The age of substitutability: What do we do when the mercury runs out? Science 191:683–689.

Handbook of Chemistry and Physics. 1962. 44th Edition. Cleveland: Chemical Rubber Co.

Herman, R., S. A. Ardekani, and J. H. Ausubel. 1989. Dematerialization. Pp. 50–69 in Technology and Environment, J. H. Ausubel and H. E. Sladovich, eds. Washington, D.C.: National Academy Press.

INFORM. 1995. Toxics Watch. New York: INFORM.

Jin, D. 1993. Optimal Strategies for Waste Management: The Ocean Option. Woods Hole, Mass.: Marine Policy Center, Woods Hole Oceanographic Institution.

Modern Plastics. 1960. 37(5).

OECD (Organization for Economic Cooperation and Development). 1991. Environmental Indicators. Paris: OECD.

OECD (Organization for Economic Cooperation and Development). 1994. Environmental Indicators. Paris: OECD.

Rathje, W., and C. Murphy. 1992. Rubbish. New York: Harper Collins.

Rogich, D. G., et al. 1993. Materials Use, Economic Growth, and the Environment. Paper presented at the International Recycling Congress and REC'93 Trade Fair. Washington, D.C.: US Bureau of Mines.

Ross, M., E. D. Larson, and R. H. Williams. 1985. Energy Demand and Material Flows in the Economy. PUCEE Report No. 193. New Jersey: Princeton University Center for Energy and Environmental Studies.

Sousa, L. J. 1992. Towards a New Materials Paradigm. Washington, D.C.: US Bureau of Mines.

Tenner, E. 1996. Why Things Bite Back: Technology and the Revenge of Unintended Consequences. New York: Knopf.

US Bureau of the Census. 1975. Historical Statistics of the United States, Colonial Times to 1970. Washington, D.C.: US Government Printing Office.

US Bureau of the Census. 1991. Statistical Abstract of the United States: 1991. Washington, D.C.: US Government Printing Office.

US Bureau of the Census. 1992. Statistical Abstract of the United States: 1992. Washington, D.C.: US Government Printing Office.

US Bureau of the Census. 1993. Statistical Abstract of the United States: 1993. Washington, D.C.: US Government Printing Office.

US Bureau of the Census. 1994. Statistical Abstract of the United States: 1994. Washington, D.C.: US Government Printing Office.

US Bureau of the Census. 1995. Statistical Abstract of the United States: 1995. Washington, D.C.: US Government Printing Office.

US Bureau of Mines. 1976 and various other years. Mineral Facts and Problems. Washington, D.C.: US Government Printing Office.

US Bureau of Mines. 1990. The New Materials Society. Vols. I–III. Washington, D.C.: US Government Printing Office.

US Congress, Office of Technology Assessment. 1992. Managing Industrial Solid Wastes from Manufacturing, Mining, Oil and Gas Production, and Utility Coal Combustion. OTA Report No. OTA-BP-O-82. Washington, D.C.: US Government Printing Office.

US Environmental Protection Agency. 1992. Characterization of Municipal Solid Waste in the United States: 1992 Update, Final Report. EPA Report No. 530-R-92-019. Washington, D.C.: US Environmental Protection Agency.

W. E. Franklin and Associates. 1990. Paper Recycling: The View to 1995, Summary Report. Prepared for the American Paper Institute, Prairie Village, Kansas.

Wards Automotive Yearbook. 1969–1995. Detroit: Wards Communications.

Wernick, I. K. 1994. Dematerialization and secondary materials recovery: A long-run perspective. Journal of the Minerals, Metals, and Materials Society 46(4):39–42.

Wernick, I. K., and J. H. Ausubel. 1995. National materials flows and the environment. Annual Review of Energy and Environment 20:463–492.

Technological Trajectories and the Human Environment. 1997.
Pp. 157–167. Washington, DC: National Academy Press.

Toward the End of Waste:
Reflections on a New Ecology of Industry

ROBERT A. FROSCH

I want to think simply and abstractly about all of industry, considered as a single block. That block takes in materials and energy, transforms them into products and wastes, and then excretes the products and wastes. At the end of their useful lives, the products may become wastes and may also be excreted, that is, "disposed of" as waste. Increasingly, we are concerned with decreasing the amount of waste to be "disposed of," and thus with changing the nature of manufacturing processes and products. In this essay I will rethink the ecology of industry as a problem in the present and will ponder the future flows of materials within and among industries. Where might we try to go? How might we try to get there?

By any measure, wastes abound in modern economies (see Wernick et al., this volume). In the United States, the material wastes from manufacturing, mining, oil and gas extraction, energy generation, and other industries currently are on the order of ten billion metric tons per year, though a large fraction of this is water. US air emissions of materials, such as the carbon in carbon dioxide, are approximately two billion metric tons per year. The end products of industry annually turn into about two hundred million tons of municipal solid waste in the United States.[1]

Traditionally, waste is whatever material is left, to be disposed of later. The emerging field of industrial ecology shifts our perspective away from the choosing of product designs and manufacturing processes independent of the problems of waste. In the newly developing view, the product and process designers try to incorporate the prevention of potential waste problems into the design process (see Frosch, 1992, 1994, 1995). Industrial ecology notes that in natural ecological systems organisms tend to evolve so that they can use any available source of useful materials or energy, dead or alive, as their food, and thus materials and energy tend to be recycled in a natural food web.

Even with the design of products and manufacturing systems to minimize waste, some waste energy and material, whether from manufacturing or from products at the end of their useful lives, will be inevitable. The second law of thermodynamics (in the simplest of terms, the impossibility of converting all the heat from a reservoir of energy into useful work) insures that there will at least be waste energy from a process; waste energy frequently appears as waste material, or it may be carried away as heat in a material. We do not know how to design processes that are perfectly economical with materials, and, in any case, this may be impossible to do.

Nevertheless, the idea of industrial ecology is that former waste materials, rather than being automatically sent for disposal, should be regarded as raw materials—useful sources of materials and energy for other industrial processes and products. Waste should be regarded more as a by-product than as waste. Indeed, as part of the design process for manufacturing end products, wastes might be designed to be useful by-products. The design optimization process would include the generation of waste, the design of waste, and the cost consequences of alternatives, for example, in reuse or disposal. This practice occurs in some parts of some industries but is not widespread.

The overall idea is to consider how the industrial system might evolve in the direction of an interconnected food web, analogous to the natural system, so that waste minimization becomes a property of the industrial system even when it is not completely a property of an individual process, plant, or industry.

I will approach the problem by trying first to abstract the essence of industrial systems in an ecological sense, subsequently considering how we might characterize and graph future states of industrial ecosystems. Next I will contemplate the potential fates for wastes and how they might be balanced. Then I will look at some of the evolutionary trends of industrial ecosystems and the properties that might mark or lead to attractive states. Finally, I will examine how we might choose policies that are likely to lead toward attractive outcomes. My approach is not to forecast but to postulate future states and sets of states, inquiring as to how they might be reached via attractive, or at least tolerable, routes. I am thinking about coupled states of industry and the environment external to industry; my scale is industry, society, and the world. When I use "we," I mean "we, the society," in whatever way a social decision may be reached to do, or not to do, something technically attractive.

ABSTRACTING THE INDUSTRIAL SYSTEM

Let us consider industry, indeed, the whole of humanity and nature, as a system of temporary stocks and flows of material and energy. Materials are seen as what they fundamentally are: elements (atoms) in the sense of the periodic table, compounds (in the sense of atoms bound together into molecules by energy embodied in chemical bonds), or mixtures of elements and compounds (perhaps several of each). In this sense, everything is a dance of elements and energy. For this discussion, energy is considered at the chemical but not the nuclear level; elements remain the elements that they are. For example, plastics—polymers— are seen less as specific materials than as collections of carbon atoms, hydrogen atoms, and some others (for example, nitrogen, oxygen, and chlorine) bound together by the energy of chemical bonds.

The processes of industry re-sort atoms into various collections or mixtures of elements bound by energy. Every step in manufacturing—product creation, product use, and product disposal—is a more or less transient event, a temporary (possibly long-lived, but temporary) use of some set of atoms and energy. (We hark back to the world of Democritus and Lucretius!) In this sense, products are just way stations in the flow of materials and energy—a temporary storage of elements and bond energy. Thus the whole sequence of events from mining or extraction to disposal is seen as a sequence of rearrangements of elements and energy.

From this point of view, reuse of materials produced as "wastes" in the course of production, or as "wastes" at the end of product life, is only *another* re-sorting of the elements and energy. The point is trivial, but the general way it puts all parts of the process into the same simple framework may suggest some useful lines of thought.

Clearly, the roles of energy and energy cost determine what may be done. Energy binds materials into compounds, but energy may also be used to take compounds apart. Energy is required to drive the processes of negentropy (i.e., those that increase order or pattern, particularly those with value to humans) that separate mixtures back into the elements and compounds from which they were mixed or that assemble atoms that have been distributed in a diluted way into collections of compounds (molecules) or other atoms. Thus, the availability and cost of energy will fundamentally determine which of these processes is economically useful for transformations and when.

At each stage in a process that produces "product" and "waste," we have some choices that determine the material forms of each, and we may link the series of choices with later sets. We postulate a universe of material/energy paths through the production, life, and dissolution of any product or set of products. We can also consider each path to be a sequence of transformations from one material/energy embodiment to another. (I generalize on a larger scale the chemical engineer's view of life.) We can view the whole of material industry as a network of such paths or transformations, connected at each end (extraction of materials and disposal of products) to the environment external to the process and product, and at places in the middle (disposal of incidental waste).

PLOTTING STATES OF THE INDUSTRIAL SYSTEM

Each network of paths or transformations may be considered a "state" of industry. Given some idea of impacts and costs of alternative paths, we could, in principle, choose among them. We would need somehow to rate the environmental impact of so complex a thing as the total industrial system. For purposes of this general discussion, I will assume that we can rate total environmental impact, and we will use only one measure or graphical axis to do so. I also assume that cost needs only one measure or axis. Cost implies dollars, but it should really be thought of as cost in the most general sense: total effort, including capital, work, energy, economic opportunity costs, and all other forms of effort, somehow calibrated in dollars. To help choose we might then create a two-dimensional graph of these states of industry by plotting cost versus environmental impact, placing each alternative state in its appropriate position.

With only one cost and impact for each state, I am implicitly considering only the total social economic cost and the total environmental cost without examining the question: Upon whom do the elements of cost and impact fall? Alternatively, we could struggle with a multidimensional space in which many axes represent different kinds of environmental impact, various kinds of costs attributed to different actors, and political and cultural variables. Such a multidimensional approach could, in principle, take account of regional differences, local politics, and other concerns. For the immediate purpose, however, I will continue to illustrate the ideas with the two dimensions of environmental impact and cost.

We must then ask about objectives. In particular, what principles should we use to choose preferred states, representing networks of paths through the sequence of material transformations? We can visualize the two-dimensional graph in which each state appears as a point, resulting in a cloud of points. Given the usual uncertainties in impact assessment and cost estimation, the "points" could represent statistical haloes, but at this level of generality, visualizing points is fine. I presume we would like to choose among states as close to the origin as possible: least cost for least impact. Various states will in fact apportion the costs and impacts to different actors, but if we were to work outward from the origin, we would be looking for solutions that are "optimum" in some useful sense, even including the politics of choosing among states near the origin on the basis of where the costs and impacts fall. However, a moment's thought about the multidimensional, more realistic possibilities and the uncertainty haloes of the points would reveal that detailed optimizing principles are far from obvious.

For some given scenarios of industrial technology and industrial organization, we might expect the possible states, the points on the graph, to have some systematic relationship or lie on a curve. For example, I would expect the attainment of extraordinarily low levels of total industrial waste or "zero waste" to be very expensive for industry; extremes are frequently costly to attain. Symmetrically, absent internalization of environmental impact costs, simply "throwing away" wastes might be expected to have high impact and low cost to industry, so these state points would be in that region of the graph. A standard economic interpretation might expect the curve connecting possible states of similar technology and industrial organization to be hyperbolic in character, like a standard demand curve. In that sense, the graph of effort versus impact can be thought of as a conventional tool of the economist.

FATES FOR WASTES

To find future states desirably located near the origin of the graph, it seems reasonable to examine those that imply both the minimization of waste in the production process and the use of wastes from various production processes across industry as input materials. Finally, we would search for "best" states by some optimization process that simultaneously considers total cost and impact. This leads us to a scenario or, rather, a set of scenarios.

Let us assume that manufacturers do an excellent job of minimizing process waste under some set of economic and regulatory pressures. Some process waste will still remain to be dealt with, as will materials available as "waste" from products arriving at the end of their useful lives. Therefore, as a baseline, let us assume a future state in which an "optimum" overall balance has been struck between limiting the creation of waste during production and reusing "wastes" that are produced as raw materials in other productive processes or that result from products at the end of their lives. This "optimum" has been chosen to lead to a minimum amount of total waste impact on the environment at minimum total cost. Some waste remains with which the overall system must deal.

The scenario must employ some policy for dealing with the remaining waste. The current policy, generally speaking, is to dispose of it: destroy it, if chemical (i.e., take it apart into simpler compounds or into elements), or "bury" it. In the general spirit of provisioning for the long run, waste for which no immediate reuse possibility exists could be stored against future need. Suppose we consider so-called Superfund sites that are "contaminated" with hazardous materials to be "filing cabinets" for potentially useful materials. Would this approach cost less or more than the current sites or systems for "disposal"? The care of potentially valuable materials may be more economical than their "disposal." Such a shift in how we view "waste sites" would require more thoughtful characterization and labeling of wastes and perhaps better technologies for packaging them.

Filing cabinet storage makes sense for many elements, particularly those with volatile market prices, and for compounds that involve elements (particularly metals) that are likely to be of future use. What to do with organic materials lacking an immediate prospect of reuse is less clear. For many such materials, storage against later reuse as chemical feedstocks might be sensible. For example, pesticides are complex organic chemicals. In the spirit of petrochemical cracking and transformation of crude oil, such a mix of organic chemicals might be a good feedstock from which to extract energy and/or derive simpler compounds suitable for chemical synthesis. Excess bond energy might be available to help power the cracking process. However, the variable stream of available wastes might make it hard to maintain a stable process. The technical challenge is then to design chemical processing systems taking into account significant variability in the stream of input materials.

Many materials might be more beneficial only as sources of energy—the energy embodied and stored in their chemical bonds. The extraction of this bond energy should be viewed as a matter of chemical processing, not just burning, and therefore as requiring the same level of process control as other chemical processing. Such control has not always been practiced with "burning" to get the chemical energy back. Incineration without energy generation as part of the process seems silly unless it is clear that the price of the power and the avoided costs of other disposal of the material do not amortize the costs of adding power generation equipment and its operation to the incinerator. The "cogeneration" alternative clearly becomes more attractive if the price of energy rises.

TRENDS AND PROPERTIES

The state described above might not cost the minimum; other states nearby in the imaginary graph might achieve nearly similar impacts for a lower total economic cost to society, presumably with a different distribution of costs among the players. Even in a highly abstract and simplified picture we see complex possibilities for choice. If we go again to more dimensions, in which various kinds of environmental and other impacts and various kinds of costs appear as axes of the display, the problem is yet harder.

Before discussing the postulated state further, some other states with resemblance to it bear description. These may or may not be in the same part of the graph with regard to costs and impacts. An earlier state from our industrial/ environmental history may be described as one in which direct manufacturing costs were minimized as a way of minimizing product cost, but environmental impacts were generally regarded as external to the industry. We have recently been in a state in which direct manufacturing costs are minimized, but regulation and, increasingly, social obligation require industry to take responsibility for disposal of waste in a way that has a low environmental impact. We now see demand for movement towards a state in which manufacturing combines cost minimization with low or zero production of waste.

The historical sequence implies next the state in which waste production during manufacturing is combined with the reuse of wastes within industry, as is the possibility of varying the materials in a product and the processes of production in order to change the nature of the waste materials, making them easier for someone to use as process input materials. Incomplete consumption of material in manufacturing and reuse is also implied, therefore leaving some waste to be disposed of at some cost.

The popular term "sustainability" presumably implies maintaining the global stock of available materials for as long as possible while at the same time preserving the general environment in a livable condition for as long as possible.[2] This latter condition should hold true for disposal and storage as well.

The availability of materials implies the small disposal—that is, the reuse—of elements that have limited availability. For this exercise I consider only Earth as available. If we make the rest of the solar system or the rest of the universe available, the principle remains the same, but the numbers and time scales change.

Preserving the environment for as long as possible against poisoning by the disposal of materials would require that we not disperse materials that harm the environment, even in concentrations so low as to not seem troublesome. Dispersal at the level of one part in a million, if continued for a thousand years, can dangerously accumulate to one part in a thousand. Although this principle omits the question of the economics, choices, and technology of future generations, it does suggest that good technologies for generating negentropy at the lowest possible direct energy cost would be welcome. Although the second law of thermodynamics tells us what the lowest energy cost must be, it does not prevent us from using technologies in which part or all of the energy cost is "free" to us, that is, where energy is used that is not otherwise directly available to us or used by us. Solar energy can be such a source. So can some or all of the energy used by organisms—for example, microorganisms, because they may extract energy from sources we do not use—provided we can figure out how to have them be our collectors.

These considerations suggest that, in general, producing *concentrated* wastes that could be useful as someone else's raw material is likely to be more interesting in the postulated state than producing *diluted* wastes. This finding reverses the wisdom with which sanitary engineers began the twentieth century, "The solution to pollution is dilution." However, it also suggests that a weakness of the alternative industrial ecosystems that include the idea of "produce but reuse" may be the small fugitive spills and emissions over long periods of time. Such a state is likely to require ever-better technology and practices for control.

Let us discuss some of the other properties of the industrial system that we have developed as our example. It will require a fairly widespread, large-scale, market-enabling information system to describe those materials available for immediate purchase or delivery and those that may be contracted for on a longer-term basis. Information on potential receivers, or buyers, of particular materials will also be needed. I am unaware of any publication, comparable to the weekly slick-paper tabloid *Chemical Marketing Weekly,* that specializes in chemical *wastes,* although that publication probably does include some materials that originate as wastes.

Beyond straightforward information leading to direct transfers of material, perhaps through brokers of materials, the brokering of more detailed adjustment possibilities would be needed. Adjustments in the products or processes of possible suppliers or users might generate "wastes" for manufacturing systems designed to use them. This process must occur for the production of particular materials now, but not much seems to happen to adjust wastes.

Specialized open-market commodity exchanges of the "pit" kind might develop in waste commodities. A computer net exchange system now reportedly operates among dealers in used auto parts, and this concept might be extended to other "waste" materials. More generally, the Internet portends cheap ways to link otherwise unconnected buyers and sellers to create markets and to search large, poorly structured databases for highly specific items.

To work for the reuse of significant quantities of materials, the system would have to be based upon realistic economics, in which the alternative processes and product materials of buyers and sellers made financial sense, including process and product costs, information and transportation costs of the various alternatives, and possible final disposal costs, whether by alteration or "land-filing." For some materials lacking realistic alternative uses, fresh incentives might make recycling economically viable. These could take forms such as taxes on their disposal in order to force reuse to be economical. The idea is that for some necessary materials, the least environmental impact might occur when a limited supply of them circulated, as opposed to flowing them through the system to "disposal."

In cases where "waste" material had no immediate use or was not available in sufficient quantity for an immediate buyer, useable public information on the contents of the filing system in landfills, with realistic systems of storage and retrieval costs, would need to be maintained. Thus, as materials came into marketable value, as happens today with copper scrap for smelter material as the market price varies, their availability would easily be known, and they could reenter the active materials market.

Both the transportation and the filing-cabinet storage systems will require considerable attention to safety, and the resulting costs would automatically be part of the trading system. Such a system would likely require a new kind of regulation for waste and hazardous-waste materials, one that recognized movement and disposal by use in an industrial process as equivalent to what is now considered final disposal. In fact, it might have to be more definitive than current regulations, which seem to view nothing as final disposal. In addition, the current tendency towards an infinite chain of liability—in which no one's liability seems to be transferable with the material, and in which liability does not die, even when the material is transformed into something else—might need somehow to be altered.

The new state of the industrial ecosystem that I have described assumes a reasonable supply of waste material even after a mature system of waste prevention is developed in the manufacturing industry. Many materials will not be internally reprocessed for reuse by the manufacturer because, for example, it may make no sense for a hard-goods manufacturer also to be a metal smelter or a maker of polymers feedstock. Waste material from sheet metal and semifabricated metal parts manufacturing will also remain available as material to be reprocessed. Not everything will be amenable to internal reprocessing.

CHOOSING POLICIES

How can we decide which policies are likely to lead in the right direction, where the end states reached would be "good" and well chosen? Nearby states are likely to resemble the state I have postulated, differing perhaps in the proportions of waste minimization and reuse, or in the specific configuration of the industrial networks, which nevertheless all have roughly the same amount of material reuse. Other states may differ in their network properties, or represent different kinds of environmental impact making up the same or a different, quantitative index of total impact, or represent different total societal costs or allocation of costs to different actors.

We must keep in mind that policies will not lead to a particular chosen state. The model of the future will be too crude, and systems do not respond by behaving exactly as policymakers desire. Actors in the system respond by doing what is in their interest, or what they perceive to be in their interest, so while systems move in a direction that the policies help determine, they tend toward being pushed in the direction of a family of possible future states rather than to some specific state determined by the set policy. We need some process for deciding which policies are likely to push the system in the direction of a set of states that would be desirable or contain mostly states that are desirable or at least acceptable.

A procedure that might work in principle would be to sample the graph of states, looking for those that are considered desirable or acceptable in terms of their position in relation to the cost/impact axes. Perhaps, if we are ambitious, we might look for states that are desirable or acceptable in terms of the finer structure belonging to other axes that we collapsed into our overall indices of cost and impact; these might include regional differences, assignment of cost to other actors, and political differences.

Given that we find a set we like, we would then consider the set of policy initiatives most likely to move industry in the direction of each state. Especially interesting are the families of policies that are common to moving in the direction of various kinds of states that are desirable or acceptable, policies that also do not appear likely to push industry in the direction of undesirable sets.

My model for policy choice among industrial ecosystems is statistical mechanics, which has developed very successfully to study systems consisting of a large number of interacting elements—particularly systems in which the large number of elements and possible interactions present an otherwise almost insuperable challenge to understanding the behavior of the whole system.

Technical questions abound about this scheme, for example, about the relationship of families of policies to the resulting state sets to which they lead; about the degree to which simple measures of state desirability reasonably represent the complexity of real world variables; and about whether the policy sets leading to generally desirable states as described by the simple measures remain robust in the face of the underlying complexity. In looking at such policy choices and deciding whether they push the right way, discussions must also examine some of the ways in which a policy set might perversely or unexpectedly push industry towards undesirable states.

The way of arriving at end states need not be by a process of long-term governmental, or large-scale, collective, detailed planning and detailed regulation. My interest is in defining states and looking at policies that may foster a move in their direction, not an attempt to get to them by compulsion. I have emphasized provision of information, freedom to contract, and profitable commitments.

Experience suggests that systems with multiple tight connections between many elements tend to be brittle and easily become unstable if a few links are broken or a single actor is removed. Systems with loose connections are much more robust, especially if the network is of single and no more than double connections. The intention here is to suggest a direction for industrial development that would be realized by standard market mechanisms with the usual relatively loose, two-party network of connections or transactions. If this development is achievable, the issue of special brittleness of the new system need not arise to any greater degree than it customarily appears in any set of market relationships of manufacturers, suppliers, and customers. Clearly, the issue of robustness would need to be considered as a new system begins to develop.

CODA

I have sought to suggest a framework for thinking about materials and their flows in the context of industrial waste, about the balancing of costs and environmental impacts in possible future states of industry, and about a method of policy examination. I feel that this approach, while abstract, contains elements that make further discussion and elaboration of its possibilities worthwhile.

I believe that ten billion or so healthy people cannot prosper on Earth without a manufacturing industry and large, complex materials flows. A simple agrarian society will not be efficient or effective enough to support likely future human numbers. Yet, vast reductions in waste seem possible if we begin to reconceive the ways we understand and operate industrial ecosystems. Jumping at solutions to particular environmental problems as they arise is easy but has not carried us nearly far enough. Little effort seems to have been given to the general, long-term, large-scale global waste problem in an analytical way. We need to stimulate more such thought.

ACKNOWLEDGMENT

I am indebted to numerous colleagues for discussions over the past several years, but would like to mention, in particular and alphabetically, Braden Allenby, Jesse Ausubel, Robert Ayres, William Clark, Nicholas Gallopoulos, Deanna Richards, and Walter Stahel.

NOTES

1. Energy wastes are also huge. Even in the most efficient economies, perhaps 10 percent of the energy extracted and generated from primary sources actually serves the end user. On the connection between material and energy wastes, see Nakićenović (this volume).

2. For a general discussion of the meaning of sustainability, see Solow (1993) and Starr (this volume).

REFERENCES

Frosch, R. A. 1992. Industrial ecology: A philosophical introduction. Proceedings of the National Academy of Sciences of the United States of America 89:800–803.
Frosch, R. A. 1994. Industrial ecology: Minimizing the impact of industrial waste. Physics Today 47(11):63–68.
Frosch, R. A. 1995. Industrial ecology: Adapting technology for a sustainable world. Environment 37(10):15–24,34–37.
Solow, R. M. 1993. An almost practical step toward sustainability. Resources Policy 19(3):162–172.

Technological Trajectories and the Human Environment. 1997.
Pp. 168–184. Washington, DC: National Academy Press.

Humans in Nature:
Toward a Physiocentric Philosophy

KLAUS MICHAEL MEYER-ABICH

What is the question to which humans are the answer? Narrowly speaking, this was the riddle the Sphinx posed to Oedipus: "What walks on four feet in the morning, two at noon, and three in the evening?" When Oedipus knew the answer—man—the beast was defeated. But thousands of years of our history, and especially the scope of environmental damage in the last few hundred, have changed the beast's appearance and posed the riddle anew. Before we give our answer, we must first understand the environmental beast's question. As an approach, in this essay I explore the cultural and conceptual history of nature in the Western tradition and the reasons and chance for a shift toward a philosophy of nature centered in nature, in *physis* rather than *anthropos*.[1]

HUMBOLDT'S DISCOVERY

Alexander von Humboldt (1769–1859), the scientist and explorer, has been called the second European discoverer of America, particularly of South America. Today, we might rather recall Humboldt as a discoverer of humanity itself in nature. Even in the present we usually take for granted that "here we are as human beings," and that some of us refer ourselves, scientifically, to the rest of the world around us. The universe then appears to be our environment, the human habitat. This approach is consistent with the anthropocentric dualism of being and having, namely, to *be* human and to *have* everything else at our disposal. Quite differently, Humboldt accepted the riddle of the Sphinx as an open question. How, then, did he look for an answer?

In his comprehensive work, *Cosmos—Outline of a Physical Description of the World,* Humboldt started with the universe as the great garden of the world (*Weltgarten*). He presented its cosmogony and cosmology, and only finally pointed out that among and with many other celestial bodies in that great garden there is also a little planet called Earth. Humboldt proceeded to describe how this planet emerged from the sun and how it continues to depend on the sun's light and warmth. Next, the elements of life—earth, water, air, and fire (or energy)— are considered with respect to their physical and chemical properties. These elements then give birth to or become alive in the biosphere so that "the geography of organic life . . . directly follows the description of the anorganic phenomena on Earth: (Humboldt, Vol. I, p. 367f),[2] with the same forces and basic substances prevailing in both spheres. Finally, humans are recognized as part of the biosphere, which itself is part of Earth and which again Humboldt points out as a particular place within the great garden of the universe.

Humboldt's answer to the riddle of the Sphinx is Copernican. He neither presumes that Earth is the center of the universe, nor that humanity is the center, but openly accepts the question of how and where we fit into the world. His answer is that we participate in the whole as part of a part of a part of it, and that we find ourselves at our place within a family of living beings, or together with others. These others are essentially *with* us, not around or for us. In accord with Humboldt I call them the co-natural world, instead of the "environment" with its unfortunate anthropocentric connotation.

Humboldt went further, and this next step may justify attributing to him the modern scientific self-recognition of humanity in nature. Thus far humanity has been identified materially as part of the biosphere, but a truly holistic[3] description must consider "nature . . . in both spheres of her being, in the material as well as in the spiritual" (Humboldt, Vol. I, p. 32). Humanity is organically equipped with reason, as a fish is with the faculty to swim. As Immanuel Kant wrote, "Reason is a gift of nature" (Kant, 1964, Vol. 6, p. A390).[4] In the study of the history of language, a preeminent tool of reason, Humboldt (Vol. II, p. 143) specifically considered humanity as a living natural whole (*lebendiges Naturganze*). In fact, he wrote that "language is . . . part of the natural history of mind" (*Naturkunde des Geistes*) (Humboldt, Vol. I, p. 383f). "The natural history of mind" sounds strange to the modern philosophical ear, but is not the perception of nature itself a natural process, so that human awareness—and that of other beings—of the conatural world and of the whole itself must be considered within the description of nature? In this sense, the first part of the second volume of Humboldt's *Cosmos* deals with the perception of nature in the poetry and visual arts of Homo sapiens. A historical outline of the scientific perception of nature follows in the second part.

Humboldt (Vol. I, p. 69) asserted that nature should not be conceived "as if mind were not included in the whole of nature" (*als wäre das Geistige nicht auch in dem Naturganzen enthalten*). Among the millions of species, nature rather has produced quite a few with faculties of language (many more have consciousness, and all, as we know, have DNA, a "syntactic language"). One has a particular awareness of the whole, so that nature recognizes herself by means of reason in the human mind. In fact, after billions of years in natural history, one of the many beings that had emerged from evolution raised its head and, in Greek antiquity, called the whole what it is: *cosmos* and *physis*. Humboldt's approach to science bears a chance to avoid the basic inconsistency of modern science as otherwise developed, that is, to comprehend the world except for a blind spot with respect to the most basic fact of that comprehension—namely, that the world includes scientists who strive to comprehend it.

The mainstream of modern science did not absorb Humboldt's Copernican insight but has only replaced geocentricism with anthropocentricism—one wrong answer for another. I state this easily, observing the environmental crisis of industrial society, but the challenge to develop a truly Copernican science still seems almost beyond human capacity. Even the theory of evolution has yet to be well understood as the basis of a natural history of the human mind. With respect to the recognition of the scientist within scientific knowledge, quantum theory in Niels Bohr's understanding seems most advanced, but the industrial economy is still based on classical physics. While the Copernican challenge to Western thought five hundred years ago evoked a development that comprises anthropocentricism as well as evolution and quantum theory, this development has impetus apart from such elements. Identifying this development further may help in considering the chances to stand up for the Copernican challenge within the next five hundred years—time scales of centuries being required for the profound penetration of major ideas into human cultures.

THE COPERNICAN TURN IN THE RENAISSANCE

The Copernican turn is usually dated 1543, the year Nicolaus Copernicus's book on the revolutions of the celestial bodies was published. In it he maintained, of course, that Earth and the other planets orbited the sun while the outer sphere remained fixed. Astronomically, this view revived Aristarchus's heliocentric system, which had been rejected for good physical reasons in antiquity. Neither Copernicus, who knew the former discussion, nor—almost one century later— Galileo had much better physical evidence in favor of the heliocentric idea than Aristarchus in the third century B.C. Newton had such evidence in his celestial mechanics, but the Copernican doctrine was no longer disputed in his time. Scientifically, the issue was not settled before 1686. That Copernicus had been largely accepted long before—mainly through Galileo's protoscientific and rhetorical exposition—shows that the Copernican issue was not fundamentally scientific. Rather, the Copernican turn was a broader cultural matter.

Galileo knew for what he fought: it was not astronomy, but the autonomy of modern man to locate himself in an open universe, no matter what the traditional authorities said. Although this claim was not presented for the first time in 1543, Copernicus's book definitively expressed the emerging spirit.

When did the Copernican turn begin? In modern times, artists usually have been most sensitive to what is emerging in history. For instance, the Impressionists were the first to paint their own painting apart from what they painted in particular, about a third to half a century before Einstein's relativity theory and Bohr's interpretation of quantum mechanics. To paint a tree as well as its paintedness, or form of perception, is basically the same idea as to consider an experimental setup as part of the physical reality of the observed object. Going back in time from when Copernicus read proofs on his deathbed, we find Leonardo da Vinci and Raphael in Florence; Antonello in Messina; Albrecht Dürer in Nürnberg; and Giovanni Bellini, Giorgione, and Titian in Venice. Looking into some of the faces in their portraits, we immediately know what happened. We see modern man at his best: open-minded and free to take up quarters in an open world.

Observing how the painters relate the saints as well as humanity to the co-natural world clarifies my meaning.[5] Jan van Eyck's *The Virgin and Child with Chancellor Rolin,* painted about 1434 in the Netherlands, provides an early example. The Virgin is painted without a halo, and the chancellor approaches her size. Kneeling before her in adoration, he is the donor of the painting. Between them we look along a river into the landscape; a bridge crosses the river, and there is a city on both sides. The Virgin and the chancellor appear on a balcony, below a roof but open to the world outside. Some people are shown who have turned their backs to the celebrities and look down the river and to the city. Mary and the chancellor seem a little too big to fit into the world shown behind them.

The formerly dominant religious paintings, which portrayed the biblical characters or later saints, had been gradually invaded by landscapes, animals, and plants in the background. Van Eyck took a further step. Man himself now came into the picture, into his own portrayal of the universe. We do not only live among other things and beings, but we begin to consider in these paintings how we put ourselves in the world, together with the sacred figures who had become the paradigms of humanity. Giovanni Bellini's *Madonna with Child,* painted in 1510 when he was eighty-two years old, illustrates how the saints are gradually indented into the world: Mary sits on the earth in a broad imaginary landscape with trees, mountains, animals, and people, holding the child. A screen behind still shields them, but only slightly, from the secular world. A little earlier Raphael had already omitted the screen in his *Alba Madonna,* showing Mary as a young woman in the country.

A little later than van Eyck, Rogier van der Weyden (1400–1464) painted almost the same scenery in *St. Luke Drawing the Virgin.* The Virgin sits on the left, St. Luke stands on the right, and again there is a river. In the city some fifteen people, engaged in various activities, can now be identified individually, in contrast to van Eyck's painting. Moreover, St. Luke, looking like a similarly educated and modern man as the chancellor, draws Mary. Formerly only the holy figures had been painted in the manner that St. Luke uses to draw Mary in the painting, but now this activity was itself painted as a process that took place in nature. Rogier painted how painters related themselves to the saints, embedding into the world the painting of the former painters who painted only the saints apart from the world. I am not suggesting that he intended this reflection, but this is what I now observe when I step back once more behind the artist who portrayed a painter at work.

When Albrecht Dürer painted his self-portrait as a twenty-seven-year-old man in 1498, he even more appeared—at least individually—on his own Copernican turn. As the art historian Erwin Panofsky (1977, p. 56) observed, this was "perhaps the first autonomous self-portrait at all," that is, painted for no other reason than self-recognition. Three years earlier Dürer had returned from his first trip to Italy, where he had met Bellini in Venice and had become conscious of himself as a painter. Dürer looks at the viewer from the left side of the painting. On the right side behind him we can identify a landscape that he had seen and painted on his trip to Venice. He thus paints himself as someone who became aware of himself when he set out into the world as a painter and related himself to the co-natural world.[6]

Leonardo da Vinci's *Ginevra de Benci* (circa 1474) provides an earlier example of a completely secular painting of a person fitting into an open world. The beautiful lady is shown with a dense, tree-like broom behind her. Beside the shrub, further in the background, Leonardo paints a lake, trees, and a distant city. This is not a portrait of a lady with a scenic background, as if the lady stood in front of a picture of a landscape. Rather, the lady fits into the landscape, participates in it, or at least is embedded into the co-natural world that is already cultivated by humanity.

How man himself takes up residence on Earth without anymore relating himself to the religious paradigms is also the topic of Giorgione, Titian's great predecessor, who died in his early thirties in the plague of 1510. *Il Tramonto,* for instance, painted probably around 1504, shows several men at a lake between mountains, a knight killing an animal, and two wanderers facing another creature in the water. Giorgione is supposed to have thought of St. George, St. Rochus, and St. Gottardus in this painting. If so, it is surprising how completely the saints have turned into human beings. Moreover, the subject seems to be nature herself. The whole scenery is animated; even the rocks seem to participate emotionally. In the background a city fits into the landscape. A similar Giorgione work, *The Tempest,* is frequently considered the beginning of modern secular Western painting.

Consciously locating ourselves in the world coincides with a growing self-esteem with respect to the religious paradigms. The entry of donors into a religious picture makes this self-esteem conspicuous. First they appear as tiny figures, adoring the saints like mice from the bottom of the picture. Then the donors enlarge until they reach the size of the saints. Palma il Vecchio even let put his hand on the shoulder of one donor, a man of his size who is place picture only a little below the *Sacra Conversazione*. It is as if man were grown up and now felt like taking the lead himself.

The Renaissance outlook of an open mind into an open world—which men and women can enter from the closed sphere of their former being to find their own place to settle down in self-esteem and confidence—spread in Europe during the sixteenth and seventeenth centuries. A learned man who spent his twenties mainly in Italy, Copernicus was sensitive to the spirit of the age. Belief in the geocentric system, or the closed world of the Middle Ages in which everybody and everything belonged somewhere, receded only in the seventeenth century. Of course, the openness of the mind, the world, and the future demanded a struggle against the old authorities, and, as far as openness was achieved, steps had to be taken on newly opened paths to make oneself at home in modern times. Some new security about one's location in the world had to replace the security implied in the geocentric system. In retrospect, this basic new security was dominantly found in the anthropocentric substitute for geocentricism. It would in turn begin to fade in the more recent transformation of the Western wealth society into a society primarily at risk from itself.

In the environmental crisis, other ideas that meet the Copernican challenge could also get their chances. For instance, a truly Copernican answer to the Copernican challenge was developed by Giordano Bruno, who was burnt in 1600 for insisting on God's infinity as a property of the world. Bruno referred to Actaeon's fate in Greek mythology (cf. Bruno, 1989, Part I, Dialogue 4). Actaeon, a hunter, was strolling around in the forest one day after a successful hunt and happened to enter into a sacred grove where Artemis (Diana), the virgin goddess, was bathing in a pond. Suddenly, in view of Artemis, Actaeon became inflamed with love for her. He was changed into a deer, from the hunter into what he hunted. As such, he was killed by his own dogs.

Bruno compared Actaeon's hunt to the search for knowledge, Artemis to nature, and her twin brother Apollo to God. Actaeon's thoughts meet their object in the things and beings of nature. As the knowing subject, he is not part of the object. Then he gains an awareness of nature herself, Artemis, the nature of being, beyond the things and beings of nature. This changes him from knowing to being known. He looks no more for others, but he is looked for himself. In view of nature herself, he is made to feel like those whom he had previously made his object, being with them instead of objectifying them. Now, by falling in love with the goddess, he has the experience of belonging to nature, like those other beings of nature. He recognizes himself in the experience of others, loving neither himself nor the others *per se* but the goddess, nature, to whom they together belong and who relates them to one another.

Could the environmental crisis have come about if we had felt ourselves how we treated others, the co-natural world? I think not. In being like others, our experience would not be restricted to having them. Science and technology have not reached that degree of maturity yet, but as researchers look more for the knowledge of maintenance instead of the knowledge of destruction, Bruno's approach may become a paradigm. If we became aware of ourselves in treating the co-natural world scientifically and technologically, a solution to this paradox might emerge.

SCIENCE FOR THE FUTURE

Francis Bacon (1561–1626) is frequently considered the first spokesman of industrial society as developed so far, as well as an anthropocentric guidepost toward environmental misery. Bacon did ignore our being part of nature. He proposed that nature must be examined like a prisoner at the bar: the scientist is the judge, and the rest of the world is exposed without pity to his art of extorting information from the indictee. This position could not be understood if Bacon had considered the judge himself as subject to the suit. Referring to Proteus, the prophetic old man of the sea in Greek legend who demonstrated his capacities to convert himself into any shape only when seized and held, Bacon (1858, p. 141) openly recommended investigating nature not free and unbound, but bound and brutalized (naturae constrictae et vexatae). Industrial society's blackout of its own belonging to nature, and its exploitation of nature as that which we are not, are essentially Baconian.

Yet Bacon had other and, in my view, better ideas that do not deserve to be abolished together with his anthropocentricism.[7] In particular, I credit Bacon with two basic suggestions on the Renaissance philosophy of nature that should be maintained beyond the environmental crisis. One is that nature must essentially be considered not for what it is but for what it is to be in the future. This differs from the Greek tradition. For Plato, knowledge referred to "ideas," and these expressed being. Others, like Heraclitus, emphasized change, but Greek philosophy did not get beyond "becoming to be" (génesis eìs ousían).

Bacon interpreted nature as production, like an artist bringing forth his work. To understand "genius and industry" (ingenium et industria), Bacon wrote, it will not do to know what the artist's material was at the outset and, apart from that, only the finished work. Instead, one ought to watch how the artist proceeds and how his work comes about. "The same is true for the observation of nature" (Bacon, 1859, section 41). What we should be interested to find out in the study of nature then is also, according to Bacon, her "ingenium et industria." This interest makes sense only when we care what nature can become.

Indeed, Bacon (1858, section 81) firmly maintains that the goal of science lies not in itself—the knowledge of being—but in the provision of new means and inventions for human life. If this is so, we then want to know what can be brought forth or produced from a given state in nature, not what it is. This does not necessarily imply that we consider the nonhuman world as only a resource for human purposes. As soon as we refrain from anthropocentrism, the new paradigm can be considered with the question: What is it—if not industrial destruction—that humanity is due to bring about in nature, to which it belongs, assuming that in the history of nature we are not supposed to leave the world as if we had not been here, which is not possible anyway? Perhaps this is the contemporary restatement of the Sphinx's question.

With respect to Renaissance consciousness, knowledge of the means and inventions that we can produce in nature promised equipment for the ways to proceed with an open mind into an open world. Bacon also used the word progress, which later narrowed an open proceeding to a particular track no longer open to changes in orientation. Bacon (1858, section 128) promised that progress in the sciences would provide the means and inventions.

The second Baconian idea I want to save from the anthropocentric fallacy is the unity of order in nature and society. This idea appears in his legal approach to scientific knowledge, which for us must not necessarily imply the use of violence that is recommended by Bacon. On the contrary, a modern understanding of a legal order is a frame for freedom and peace, excluding violence. Bringing this approach to bear in science and technology at today's level of legal and political culture could prove a great step in Copernicanism.

Moreover, Bacon had a coherent idea of political order in society and nature. The most primitive political ambition, he thought, is to come to power in one's own country. Having achieved this—as Bacon himself had—the next and more noble ambition is to bring one's own country into power internationally. Bacon's idea—which he conceived when he turned to science after losing office—was to extend power to a third level on top of the two political levels of human domination within humanity, namely, the domination of humanity over nature (Bacon , 1858, section 129). If we take this to mean human domination in nature, considering our own participation in nature, and if we interpret domination more broadly as political order, Bacon's idea means that the social order of life is to be embedded in a more comprehensive order of nature in which humanity relates itself to the co-natural world. The laws of nature do not then essentially differ from the laws of justice. Both belong to a more comprehensive common order. Because the basic issue in the environmental crisis is precisely to embed societal order— especially economic production and consumption—into the order of nature, Bacon's approach could offer a chance to fit together what has fallen apart.

Baconians may be unhappy to give up anthropocentricism and to reduce domination to its objective—order—but my question is, to what extent can Bacon's views, which have led us into the environmental crisis, also lead us out of it? Relating the laws of justice to the laws of nature meets the objection that we should not commit naturalistic fallacies, i.e., draw conclusions from what "is" as to what "ought to be." But this rule applies only insofar as the statements about nature remain strictly descriptive, and this is not the case in Bacon's concept of nature as something to be. There is no fallacy in a prescriptive inference from a prescriptive description. Bacon's legal approach in establishing the rules on how to produce in nature shows that the dichotomy between the two kinds of laws was far from what he intended. In a letter to the king Bacon argued explicitly that the rules of nature and the true rules of politics are related:

> I do not find it strange . . . that when Heraclitus . . . had set forth a certain book which is not now extant, many men took it for a discourse of nature, and many others took it for a treatise of policy and matter of estate. For there is a great affinity and consent between the rules of nature, and the true rules of policy: the one being nothing else but an order in the government of the world, and the other an order in the government of an estate (Bacon, 1868, p. 90).

The observation of this affinity was certainly intended to recommend Bacon, the scientist, as an advisor to the king. In any case, the affinity between the two orders becomes a challenge to contemporary politics not generally expected from Bacon. But nowadays it is a common experience that politics becomes very unpolitical when exclusively concerned with human affairs, if not only with those of the politicians; at the same time science and technology again and again prove to be the most relevant political activities in industrial society. For instance, no modern foreign minister has influenced international relations to the extent that Otto Hahn, Fritz Strassmann, and Lise Meitner did with their discovery of nuclear fission, and no minister of economics or labor has ever been as effective in these fields as the applied physicists of Silicon Valley. This would not have surprised Bacon, and he can remind us that the two orders are interlinked.

To overcome the barrier, it must be observed as a political fact that industrial society still considers science as describing what nature is, not what it is to be, and still considers the experiment as a means to find out what nature is. Both assumptions are essentially wrong in a philosophical analysis. Rather, the rules of nature as conceived in modern science are a canon for proceeding when something is to be produced; and the experiment confirms that the rules of production of some effect have been made available.

Simple as these seem, the error in modern self-awareness of science is tough. It is always hard to understand why certain mistakes seem indispensable in one's own self-appreciation, individually as well as socially. With respect to the political character of science and technology, I believe that, because the goals of science and technology have become doubtful in many fields, industrial society does not want to be reminded that these not only provide the means to go somewhere but are themselves moving ahead. Science itself proves to be not as scientific as its results, so to speak. In such a situation a historical anamnesis of what had once been the goals might prove helpful, and this involves reconsidering the idea of progress.

PROGRESS INTO AN OPEN WORLD

Progress originally meant simply to start, to set out from fixed relations. The European crusaders set out for the Holy Land, and the European explorers set out around the world. Science set out to continue inside nature what the explorers had started externally. Like Columbus sailing for India, Bacon had a definite idea of how to venture into the Renaissance open world, namely, by means of progress in science and technology, and he also conceived where this progress should lead. The true end of knowledge, he declared, "is a restitution and reinvesting (in great part) of man to the sovereignty and power (for whensoever he shall be able to call the creatures by their true names he shall again command them) which he had in his first state of creation" (Bacon, 1859, p. 222)

Bacon's "first state of creation" alludes to a paradise as conceived in the Jewish Old Testament, where man was entrusted with dominion over nature in the name of the creator and was instructed to name the other creatures. This power was restricted after the Fall when man had to work for his own. To restore human sovereignty, as in paradise, by means of science and technology thus means that the end of knowledge is a compensation for the loss that we suffered by eating from the tree of knowledge. Progress was conceived as a dodge back into paradise, which would even include the restitution of immortality: "And to speak plainly and clearly," Bacon continued, "it is a discovery of all operations and possibilities of operations from immortality (if it were possible) to the meanest mechanical practice" (Bacon, 1859, p. 222). In this paradise, moreover, the Lord was no longer required to rule, because human inventions could be considered as "new creations" (*novae creationes*), by man stepping into God's shoes. Technological man became deified in humanity (*hominem homini Deum esse*) (Bacon, 1858, section 129), especially with respect to those peoples (savages, developing countries) who had not yet entered the occidental path of progress back into paradise.

The idea of progress to paradise thus entered into our cultural history. At the outset there was just the Renaissance openness to "set out" in itself, fresh and clean like a morning—whatever the day brings, we should never forget that in any case we are lucky to be on the way. At least, that is the way I feel about the Renaissance outset. As Leibniz, the philosopher of perfection, put it, "A certain uneasiness in longing for the good together with a continuous and uninterrupted progress to the greatest goods is even better than to possess the good" (*un progres continuel et non interrompu à des plus grands biens*) (Leibniz, 1959, Chapter 21, section 36). Appetite feels better than satiety, as long as one is not starving. Apart from the happiness of being on the way in itself, Leibniz to some extent also considered where this progress might lead. He could imagine that humanity in time might reach a greater perfection than we are now able to imagine, but this was not his main concern (Leibniz, 1985, section 341). Rather, his essential idea with respect to progress was that striving for perfection is itself an element of perfection. "Without perpetual progress and novelty there is neither thought nor pleasure" (Leibniz, 1967, p. 101).

Progress also became a political idea. Understandably, it was turned against the traditional authorities. Soon enough, progress was also used to build the self-esteem of some with respect to others. French poet Charles Perrault (1628–1703), who gave classic form to many children's tales, felt compelled to prove again and again the superiority of modern times to antiquity. This quarrel was unnecessary because the point could be settled by observing, as his countryman Pascal did, that humanity can learn in the way that an individual learns. Although the moderns may know many things better than the ancients, a grown-up is not in every respect better than the child that he or she was before.

Politically more detrimental than presumptuous attitudes in regard to antiquity was the emergence of Eurocentrism. Feeling progress and that progress should lead to a particular kind of perfection made hard the appreciation of other people's ways of life. For instance, the so-called savages were, according to European standards, not as advanced as the Europeans. Those who thought, as the social philosopher Abbé St. Pierre (1658–1743) did, that humanity as a whole was on the road to (occidental!) reason could not help but observe different degrees of advancement (in Europeanism) when they looked around the world. The French economist and administrator Turgot (1727–1781) even drew a map of progress that showed Europe favorably in front of the underdeveloped savages as well as the stagnant Chinese. And Voltaire comforted the Brazilians in considering them as simply not yet fully developed to humanity; some day they would also have their Newtons and Lockes, though somewhat behind the Europeans (Voltaire, 1786, p. 271). These views sound familiar, naive, and superstitious in light of the failures of twentieth-century development policies intended to let developing countries "catch up" with the industrialized ones (see Grübler, this volume). We see that Eurocentrism is not only a political issue but is rooted in our modern consciousness. The depth of the roots may explain why occidental rationality seems even more overwhelming for other cultures than the political and economic power of the industrialized countries.

To what extent are environmental and other failures rooted in the original approach itself? What was meant and what went wrong can be distinguished in the philosophy of progress of the mathematician, social theorist, and political leader the Marquis de Condorcet (1743–1794).[8] Wrapped in the spirit of the French Enlightenment, Condorcet believed in the perfectibility of humanity. This idea, held in common by Marxists and, to various degrees, by those in the biotechnology endeavor, rings several bells when it now arises. Condorcet's point, however, was that it is a mistake "to consider man as being shaped by the actually prevailing state of civilization as natural" (Condorcet, 1976, p. 78). Instead, natural should be defined as what is to be and what is possibly to come if we set out for the better. To give an example that Condorcet himself used: For the time being, superiority is generally to the advantage of those who are superior and to the disadvantage of those who are inferior. Many people believe that this is "natural," and in saying so believe that such is life, that it must be faced and will never be different since what is natural must be accepted as such.

Condorcet might not disagree with this understanding of "natural," but he refused to accept as "natural" the idea that the use of superiority was to the advantage of only the superior. And should we not agree with him? Will human affairs ever change for the better if we simply accept their shortcomings—no matter how common—as natural and therefore inevitable? Condorcet did not refrain from the hope that if a natural state is to come, "superiority will turn into an advantage even for those who do not have it" (Condorcet, 1976, p. 203). Is this not a goal that we should accept?

To consider nature—like Bacon and unlike industrial society—as something to come was a common idea to the philosophers of the Enlightenment and of the French Revolution. "In the name of nature they requested beings and military commanders, officials and priests to respect human life" (Condorcet, 1976, p. 159). This confidence in nature was based on the scientific experience that the universe had proved to be in good order. "Why should this principle less apply to the development of man's intellectual and moral capacities than to the other processes in nature?" (Condorcet, 1976, p. 193). Not only the laws of the nonhuman world but also "human rights are written in the book of nature" (Condorcet, 1976, p. 121). Nature has planted the seed for better social relations into our hearts, waiting only for enlightenment and freedom to develop (Condorcet, 1976, p. 212).

Apart from the optimism with respect to reason and revolution, I agree with Condorcet's position. We must be careful not to lose hope when we abandon optimism. Condorcet's optimism, however, was fatal. When Condorcet refers to breeding better animals, for instance, as a parallel to human perfectibility by education, and when he assumed that the human life span would continuously increase (Condorcet, 1976, p. 219ff), he is on a straight line from Francis Bacon—even from Roger Bacon in the thirteenth century—to modern biotechnology. Condorcet did thoroughly condemn genocide by European colonialism and proposed reparations in the form of access for underdeveloped peoples to participate in European Enlightenment. Well-meant as this notion was, spiritual imperialism is not so far from military and economic imperialism as Condorcet seems to have thought.

Condorcet's gasping totalitarian optimism may reflect an imminent sense of his personal fate. He had supported the French Revolution but opposed the death penalty and consequently the regicide. Persecuted, Condorcet wrote his epochal review of human progress in the span of a few weeks, then he left his shelter and was found frozen to death in early 1794. Considering his technocratic phantasms in the context of his personal situation, they seem like morbid dreams. So do many modern technological developments, civil and military, which are equally as high tech as Condorcet's visions.

Where then will we find the breath for a Copernican existence in nature to come, toward which the Renaissance spirit set out? I will conclude this essay by reapproaching Humboldt, now from the eighteenth century, by backing his Copernicanism with Immanuel Kant's idea about nature's intention in human history.

PROCEEDING TO NATURE IN HISTORY

In his *General Theory of Heaven* (1755), Kant challenged Newton's compromise with Bishop Bentley. In this compromise, theology left the explanation of the planetary movements to science, while science accepted theological competence to explain how the planetary system had come into being. Kant, however, not only edged science into the domain that Newton had left to theology but based his theory on a different concept of nature. While Newton and Bentley considered nature, as Kant put it, as "an obstinate subject" that must be forced into order, Kant (1960, p. A194) conceived the beings of nature to be akin by a common origin and by themselves apt to an orderly constitution. It is true that, in the West, before modern science the nonhuman world had not been expected to be deliberately ordered. In fact, even in heaven the planets had been termed "the wanderers," moving back and forth irregularly, as it seemed. The discovery of this formerly hidden order, however, convinced Kant that nature was not the obstinate subject it had been considered, and in his philosophy he ventured to explain how its order is constituted.

Kant did not doubt that as humans are a part of nature, human history equally is. Modern historians, however, tend to miss this point. After the unexpected recognition of the orderly structure of the nonhuman world, the challenge for Kant became whether perhaps even in human history—apparently the realm of unreason and arbitrariness—a corresponding order could be found. To consider one and the same nature to embrace the human and the nonhuman world aligned with traditional thought from Plato to Bacon and then on to the philosophers who looked for human rights written in the book of nature. But the success of modern science had become a particular test for that unity. After publishing the first edition of the *Critique of Pure Reason* (1781), Kant set out to accept the challenge in his *Idea of a General History in a Cosmopolitan Intention* (1784), looking for an intention of nature (*Naturabsicht*) in the apparently erratic course of human history (Kant, 1964, p. A387). The idea of nature in Kant's philosophy of history does not fit together with deterministic nature as opposed to human freedom in the first two *Critiques*. Yet it serves as Kant's turning point from these to the *Critique of Judgement* and to his post-critical, more "organic" philosophy. It was Johann Gottfried Herder who provided the inspiration for Kant's decisive change with his *Ideas in the Philosophy of History of Mankind* (1784–1791). Kant's aggressive review of Herder's *Ideas* seems to indicate that Kant did not appreciate that his former student had been ahead of him here.

The search for an intention of nature in human history does not presume nature to be intentional like humans. Kant accepted epigenesis rather than preformation. The former assumed the existence not only of actual properties but also of dispositions to become what has not yet been developed, i.e., the existence or reality of possibilities. For humans, Kant deemed the natural dispositions to be reason, freedom, sociability, selfishness, and a moral sense, though this last was considered rather crude.

With respect to the five dispositions, Kant's "idea" of 1784 was not that they would develop by themselves, like the different branches of a tree, but that humanity is developing by means of the conflict between sociability and selfishness. As he put it: "Man does not like his companions, but he cannot escape them either" (Kant, 1964, p. A393). Driven by vanity and avarice, he competes with them, and this competition produces culture as the specific human contribution to the evolution of nature. In contrast to Adam Smith's shallow optimism, Kant did not assume that the individual pursuit of happiness would produce the common good. With respect to happiness, Kant (1964, p. A393) wondered if humans might be better off if they lived modestly like sheep, but as such they would not fill humanity's ecological niche or the space which is left open for humanity in nature, namely "reasonable Nature" (*vernünftige Natur*). Nature fitted us with reason and the freedom to use it (Kant, 1964, p. A390), not to become happy, but to develop this particular disposition by means of the conflict between the individual and society. The outcome, Kant suggested, should be a civil society in a legal state of nature, this being "intended" by nature in cosmopolitan human history. But this is only our *chance* in the history of nature, and it is up to us to seize it.

Kant's philosophy of history lacks totalitarian optimism as well as the presumptuousness of the philosophers of perfection by progress, and at the same time is truly Copernican. The question is how we locate ourselves in nature, or what is our ecological niche, and not what the rest of the world has to offer to please us. In fact, in his later philosophy Kant's conception of nature tends to become non-anthropocentric. There is no being in nature, Kant (1957, p. B382) maintained in his *Critique of Judgement,* that may presume to be the purpose of everything. Rather, those who appreciate the beauty of nature admire and love the other living beings—including the plants—even if it is to their individual disadvantage (Kant, 1957, p. B166f), because a circle exists of "mutual dependency of all beings, man not being exempted" (Kant, 1936, p. 570).

To the challenge of whether in human history an order corresponds to the one in nature outside humanity, Kant answered that in spite of present disorder there is the chance of an order of natural law that may be reached in the future. In accordance with the Renaissance tradition, nature is again conceived of as what is to be. Along the same lines as and following Leibniz (1992, section 2, p. 275), Kant (1957, pp. B77, 270) conceived of the co-natural world as art, and the reasonability of its order as art's intentionality. Correspondingly, a work of art is as perfect as it looks natural so that the creative power of nature is effective in the productive capacity of the artist, too. Art is the evolution of nature in freedom, which of course includes "abstract" as well as "concrete" art as far as its naturality goes (Kant, 1957, p. B200).

To conceive of nature as what is to be intended applies equally to the "laws of nature" in science, once they are—as mentioned earlier—identified as a canon of rules for how human intentions in nature can be implemented. Kant provides the paradigm, that something is known as soon as it can be deliberately produced, in a famous formulation in the *Critique of Judgement*: In science we want to find out "what we are able so to subject to our observation or experiment that we could ourselves produce it like nature, or at least produce it according to similar laws. For we have complete insight only into what we can make and accomplish according to our conceptions." (Kant, 1952, p. B309). In other words, being is known as being produced. Scientific knowledge is practical knowledge. Science does not simply deal with matters of fact but with matters of effecting facts.

Kant's philosophy of history may be considered a *"physiodicee,"* a justification of nature for bothering us with enormous confusion compared to the reasonability of the co-natural world. Nature is justified by the chance humans have finally to create peace in a legal order of natural rights.

CONCLUSION

This *tour d'horizon* of the cultural and conceptual history of nature in the West leads to the view that the human challenge is to justify how we proceed to locate ourselves in an open cosmos, which centers neither on Earth nor on humanity but on its own nature, nature herself. Oedipus must broaden his answer.

Our present state is to live without peace, to live in violence, alienation, and disorder. To endure this state and to use nature's gifts of human disposition, we have the chance in the future to recover what has been lost. We are actually living in evil, but better circumstances and ultimately peace may emerge from this intermediate state between nature lost and nature recovered.

Apparently, we need the industrial economy to treat problems we would not have without the industrial economy. More generally, we will need science and technology to treat problems that we would not have without science and technology. And we need to diffuse a new understanding of nature, including our own nature, in order to drive our science.

Finally, the environmental crisis reminds us that peace is not a matter of humanity being a closed society, but rather ought to be found in siting ourselves in the whole of nature. We might succeed in so doing within the next five hundred years.

NOTES

1. Many questions must be left open in this survey. A more comprehensive picture will appear in K. M. Meyer-Abich (in Press).

2. Quotations from non-English texts are given in the author's translation, except for *The Critique of Judgement* on page 182, this volume.

3. The term "holistic" is used in its philosophical sense within the philosophy of nature, as developed by Smuts (1926) and A. Meyer-Abich (1934). Contemporary discussions about holism could benefit from using the term in a more specific sense.

4. Kant quotations refer to the pagination of the first (A) or second (B) German edition of the particular paper or book.

5. A thorough study of the Renaissance spirit as expressed in portraits has been presented by Boehm (1985).

6. I am indebted to Richard Hoppe-Sailer and Frank Fehrenbach at my institute for a stimulating discussion of Copernicanism in Renaissance printing, and to Richard Hoppe-Sailer particularly for pointing out Dürer's self-portrait to me.

7. The idea of a non-anthropocentric Baconian science has been put forward by Krohm (1994).

8. On responses to utopian views, see Kates (this volume).

REFERENCES

Bacon, F. 1858. Novum Organum, The Works of Francis Bacon. Vol. I. J. Spedding, R. L. Ellis, and D. D. Heath, eds. London: Longman.

Bacon, F. 1859. Of the Interpretation of Nature: The Works of Francis Bacon. Vol. II. J. Spedding, R. L. Ellis, and D. D. Heath, eds. London: Longman.

Bacon, F. 1868. The Letters and the Life of Francis Bacon. Vol. III. J. Spedding, ed. London: Longman.

Boehm, G. 1985. Bildnis und Individuum—Über den Ursprung der Portraitmalerei in der italienischen Renaissance. Munich: Prestel.

Bruno, G. 1989 (1585). Eroici Furori—Von den heroischen Leidenschaften, Christiane Bacmeister, ed. Hamburg: Felix Meiner Verlag.

Condorcet, Marquis de. 1976 (1795). Esquisse d'un tableau historique des progrès de l'esprit humain—Entwurf einer historischen Darstellung der Fortschritte des menschlichen Geistes, W. Alff, ed. Frankfurt/Main: Suhrkamp.

Humboldt, A. von. 1845–1862. Kosmos. Entwurf einer physischen Weltbeschreibung. Five Volumes. Stuttgart/Tübingen: Cotta.

Kant, I. 1936. Kant's handschriftlicher Nachlaß. Vol. VIII: Opus postumum. Erste Hälfte (Convolut I bis VI). Kant's gesammelte Schriften Vol. XXI. Preußische Akademie der Wissenschaften, ed. Berlin/Leipzig: Walter de Gruyter & Co. Reprinted 1973.

Kant, I. 1952. The Critique of Judgement. Translated by J. C. Meredith. Oxford, England: Oxford University Press.

Kant, I. 1957 (1790). Kritik der Urteilskraft. Werke in sechs Bänden. Vol. 5. W. Weischedel, ed. Darmstadt: Wissenschaftliche Buchgesellschaft.

Kant, I. 1960 (1755). Allgemeine Naturgeschichte und Theorie des Himmels. . . Werke in sechs Bänden. Vol. 1. W. Weischedel, ed. Darmstadt: Wissenschaftliche Buchgesellschaft.

Kant, I. 1964 (1784). Idee zu einer allgemeinen Geschichte in weltbürgerlicher Absicht (Idea of a General History in a Cosmopolitan Intention). Werke in sechs Bänden. W. Weischedel, ed. Darmstadt: Wissenschaftliche Buchgesellschaft.

Krohm, W. 1994. Die Natur als Labyrinth, die Erkenntnis als Inquisition, das Handeln als Macht—Bacons Philosophie der Naturerkenntnis betrachtet in ihren Metaphern. In Naturauffassungen in Philosophie, Wissenschaft, Technik; L. Schäfer and E. Ströker, eds. Freiburg/Munich: K. Alber Verlag.

Leibniz, G. W. 1959. Nouveaux essais sur l'entendement humain—Neue Abhandlungen über den menschlichen Verstand. In Philosophische Schriften. Vol III/1. W. von Engelhardt and H. H. Holz, eds. Darmstadt: Wissenschaftliche Buchgesellschaft.

Leibniz, G. W. 1967. Confessio Philosophi—Ein Dialog, von Otto Saame, ed. Frankfurt/Main: Klostermann.

Leibniz, G. W. 1985. Essais de théodicée—Die Theodizee. In Philosophische Schriften. Vol. II/2. H. Herring, ed. Darmstadt: Wissenschaftliche Buchgesellschaft.

Leibniz, G. W. 1992. De ipsa natura sive de vi insita actionibusque creaturarum—Über die Natur an sich oder über die den erschaffenen Dingen innewohnende Kraft und Tätigkeit. In Philosophische Schriften. Vol. IV. H. Herring, ed. Darmstadt: Wissenschaftliche Buchgesellschaft.

Meyer-Abich, A. 1934. Ideen und Ideale der biologischen Erkenntnis—Beiträge zur Theorie und Geschichte der biologischen Ideologien. Leipzig: J. A. Barth.

Meyer-Abich, K. M. In press. Praktische Naturphilosophie. Munich: C. H. Beck.

Panofsky, E. 1977 (1943). Das Leben und die Kunst Albrecht Dürers. Munich: Rogner und Bernhard.

Smuts, J. C. 1926. Holism and Evolution. London: Macmillan.

Voltaire. 1786. VII Entretien: Que l'Europe moderne vaut mieux que l'Europe ancienne. In Oevres complètes de Voltaire. Tome 36: Dialogues et entretiens philosophiques. Gotha: Ch.-G. Ettinger.

Technological Trajectories and the Human Environment. 1997.
Pp. 185–198. Washington, DC: National Academy Press.

Sustaining the Human Environment:
The Next Two Hundred Years

CHAUNCEY STARR

It is surely presumptuous to look into the two-hundred-year future of this changing world. Yet the questions we pose today about sustaining the world's habitability, the environment, and the quality of life of its human population force us to stretch our thinking in time. A society takes a half century to incorporate major technological changes into basic systems such as energy and transport, perhaps a century or more to modify substantially its cultural values, and many centuries to reconcile historically embedded ethnic and religious differences. A two-century scenario may provoke useful thoughts on the likely implications of present global trajectories and apparent choices.

With such ponderous response times, today's societal institutions strain to accommodate the pressures arising from diverse forces. The global population has doubled in forty years and may double again in seventy-five more. Global economic output doubles about every thirty years, inevitably with an increased demand on our natural resources and a greater impact on the environment. New technologies that significantly transform goods and services now seem to appear roughly every twenty years. More slowly, governments and cultures adjust and restructure.

The visible consequences of this mismatch in the dynamics of forced change and social restructuring have been highlighted by the various environmental movements. Environmentalists emphasize both today's costs of growth and the dismal implications for the future. How habitable will Earth be? Can ecosystems endure the projected pressures? What will be the resultant quality of life for the world's human population?

The popular, gloomy response to these questions discounts the continuing positive contributions of science and technology, which have created a large share of today's global resources and life-style options. Many resources currently available would not be accessible or even recognized with the technology of a century or two ago. It is reasonable to ask, therefore, what might new technology contribute to our perceptions of global habitability, sustainability, and quality of life? I will briefly comment on these three concepts; look at the likely numbers of people, their food, water, and energy; and then discuss where they may live in a scientifically and technologically dynamic setting.

HABITABILITY

In its conventional sense, habitability implies: (1) the availability of the means for human survival, namely, food, water, and shelter from climate extremes; (2) physical security from the threats of humanity and nature; (3) social security and stability in human relationships; and (4) a foreseeable continuum of the above. The majority of the world's population has long considered these goals adequate, even though they exceed the reach of many today. A few also consider amenities that make life pleasurable a sophisticated necessity.

The expansion of "habitability" to include also the preservation of ecosystems apart from the value they provide to people assumes an answer to the philosophical question about the equitable treatment of all living things, of which humanity is but one (see Meyer-Abich, this volume). This essay will deal only with an anthropocentric view, but even that one-sided perspective obviously requires a sustaining relationship between humanity and the rest of nature.

SUSTAINABILITY

Assuming humanity and nature are separable entities, the environmentalist's usual definition of sustainability is development that consumes nature's dividends without impairing nature's capital. This ideal cannot be realized because, without humans, nature's dividends are already fully reinvested in the maintenance of nature's capital. Obviously all human activity intrudes and makes demands on nature's resources, both its dividends and its capital. Increasing intrusion seems inevitable with an increasing global population. Even if the use of dividends and capital could be stabilized, it would imply a zero-sum game of distributing fixed natural resources among people's regional demands for habitability and life-style. A dynamic, turbulent, competitive, and growing world population will likely find such a definition of sustainability too confining. Governments, including democracies such as the United States, so far show no success in achieving politically accepted zero-sum stability in national planning.

What would be a more viable concept of sustainability? Could it be the minimizing of progressive environmental degradation? Or the preserving of existing biotic resources for future populations? A case can be made for each.

Specific examples of their merit for environmental well-being abound, and their proponents use them to illustrate the urgency of action to achieve their view of sustainability.

Interestingly, these somewhat diverse concepts share roots whose strength is rarely examined. They all start with the finiteness of the planet. It therefore follows that the planet's natural resources must also be finite. It is said these are being consumed at an expanding rate as the world's population and economy grow. Thus, Earth's resources of minerals, water, air, and biota will be depleted by the unrestrained increase in people's activities, and if this increase is allowed to continue unmodified, catastrophes will eventually destroy any sustainable symbiosis of people and the rest of nature. Finally, the confluence of the exponential trends of population growth and resource depletion could be globally devastating.

This qualitative logic leads to the common conclusion that the present generation of decisionmakers should act responsibly through their various governments to mandate the constraints on people's behavior necessary to avoid the pending catastrophes. So, as this reasoning goes, unless we act quickly and stringently, our future generations will successively face rapid deterioration in their quality of life. And, of course, our ethical responsibility to protect our children, and their children, requires sacrifice today for this purpose. These are powerful rhetorical arguments, and their simplicity leads to a standard manifesto applicable to all the presumed evils.

This apparently logical sequence depicting environmental threats routinely provides the preamble to the massive literature on public policy relating to the environment and its sustainability. The absence of credible, quantitative characterizations of Earth's resources likely to be available for coming centuries, which may be compared with the range of future global demand, makes the arguments more compelling. The evolving availability of the resources and the efficiency of their use globally are usually neglected. Also absent is the potential for restoration of fauna and flora on a regional basis, such as through forestry. We are thus left with a vision of a likely worldwide degradation in the quality of life, though both timing and magnitude remain uncertain. The many specific examples of degradation that we see today reinforce this vision. We do not know whether these are significant early indicators of global trends or local failures to be expected in a turbulent world.

Fundamentally, these arguments provide the moral stiffening for much of the political activism aimed at persuading government to mandate actions to protect nature's future. The demand for centralized decision-making to limit the role of markets in which individual self-interest might dominate often accompanies this view. Implicit is a distrust of lay judgment in the marketplace of economic, technical, and social options. This distrust arises in part from a fear that the public reacts only when threats become highly visible or advanced, and that delayed action will fail to protect the environment. Such an environmentalist mixture of assumed prescience and virtuous political autocracy has provoked much controversy in scientific circles, as well as some popular and political backlash. In view of the real uncertainties in the science underlying most long-range environmental concerns and the intense competition among pressing social needs, this debate will likely persist.

If one seeks compatibility with the agenda of most of the world—in which people have primacy—the appropriate concept of sustainability emphasizes the efficient use of all natural resources while seeking simultaneously to minimize avoidable degradation. The globally overriding societal concern is the long-term improvement in the quality of life. Such a broad dynamic concept makes the power of the human brain the crucial renewable resource, to be applied to the management and development of natural resources necessary for human betterment.

This last view of sustainability recognizes practically the revealed priorities of most societies worldwide. It implies a major role for growing technical skills in resource development, management, and use. Many current resources result directly from technical skill, for example, photovoltaic energy, deep oil, desalinated water, and synthetic materials.

QUALITY OF LIFE

Quality of life for the world's people incorporates both the habitability concerns about fitness for life and the sustainability concerns about balancing present and future well-being. The concepts of habitability and quality of life are merged at the outset and can separate only after the habitability needs are met. The concept of a societal quality of life is highly subjective and personal, although a few criteria are generally accepted. Often cited are a low level of infant mortality and long life expectancy—the beginning and the end of the human span. These aggregate the effects of many social and environmental factors, each of which deserves public attention because of the high priority of health.

Another generally accepted criterion is education, which is roughly indicated by the literacy rate of population groups. This rate perhaps tells more about societal priorities, egalitarianism, and economic status. Literacy opens the door to the vast knowledge resources of the world and has been a requirement for a modern society. However, it is arrogant to presume that illiteracy necessarily leads to dissatisfaction with the quality of life, as is demonstrated by the happy members of many non-literate, isolated groups. It is their interaction with modern societies that creates the desire for literacy as an important component of their quality of life.

Rarely measured, but often mentioned, are the needs of the psyche—for freedom in all its aspects, choice in life-style, the beauty of nature, the arts, and the pleasures of recreation. This complex of individual priorities matters more as the other quality-of-life needs are met. This situation describes about a fourth of the world's population, particularly its more affluent members who can more easily afford a long-range view. It is the group that most visibly and vocally cares about future global trends. It is also the group that most willingly accepts the marginal sacrifices and constraints suggested for mitigation of the undesirable and uncertain outcomes that might develop in the coming centuries, in the belief that the constraints would help.

THE NEXT TWO HUNDRED YEARS

For decades the debate over the urgency and content of policies intended to save the world from future devastation has been framed primarily by doomsayers with the rhetorical certitudes suggested above. I believe it is important to temper such prognostications with a touch of reality. Societies react to threats with a combination of actions to reduce their probability and severity, usually very slowly. Premature or unwise actions in anticipating threats may themselves lead to catastrophes. For example, the obsolescent Maginot Line, designed to protect against German tactics and equipment of World War I, left France defenseless (yet confident) at the start of World War II. Policies and investments are rapidly outmoded, rendered ineffective by changing systems, and may actually inhibit the best responses provided by new technology. Nevertheless, early preventive actions with today's technology sometimes seem appropriate when threats become evident. How do we gauge our timing and strategies when all we have to go on is the current information base? This is the classic problem of decision-making under uncertainty, and it is interesting to approach it as such.

Let us now consider a plausible global development for the next two hundred years on the basis of today's knowledge, hindsight, and foresight. Hindsight teaches us not to be too confident. A recent book looked at 1993 as it was foreseen in 1893 (Walter, 1992). Except for a few forecasts, such as the growth of the telephone and cities, the great minds of 1893 failed to envisage 1993. Important technological, economic, and political changes were mostly unanticipated. Perhaps we can do no better today, but we should try, using this past century's experience, to foresee the world realistically rather than as we would like it to be.

Population

First, consider human population. A thousand years ago the global population was in a period of rough stability at about three hundred million, one person for every twenty today. By the seventeenth century, for many reasons including the onset of productive technologies, human numbers started to grow with a doubling time of about two hundred years. A century ago, global population growth had a doubling time of one hundred years. The rate peaked around 1970 at about thirty-five years. By the middle of the next century, the United Nations projects a global population of more than ten billion. Even if a global fertility reduction program instituted today could quickly reach and maintain a doubling time of seventy-five years, the population would increase sixfold to thirty-six billion by the year 2200. Some social or Malthusian limits seem likely to enforce stabilization in advance of this overwhelming number, although no convincing logic exists for a particular lower number. I judge hopefully that by the year 2200 the world's population might level off at three times the present, that is, about fifteen billion. Most of the intervening growth will be in the less-developed countries, where children are needed for cheap labor. Population growth has generally slowed with economic growth, so bettering the economies of the less-developed countries may hold the global population to fifteen billion. Can we provide food, water, and energy, the basic physical needs, for such a population?

Food

No technical or resource obstacle inhibits a threefold or more increase in the world's food production. In the industrial countries, only a small fraction of the population is needed to produce the food, and about 10 percent is involved in its processing and distribution. The work force required is thus not a limitation. At present, more than half the food grown in less-developed regions is lost through spoilage, so the delivered food supply could be roughly doubled by using modern refrigerated transport and preservation techniques such as canning, dehydration, freezing, and irradiation. Doubling arable land globally is not a limitation if reliable water for irrigation is available. In fact, with irrigation, the present agricultural lands would be sufficient to double the world's food production. The crucial issues are the availability of water and energy to support the growing, processing, and distribution systems, and perhaps unwanted side effects from intensive production (see Waggoner, this volume).

Water

Water is both the world's most valuable and most wasted resource. Globally abundant but unevenly distributed, everyone understands its essentiality, but few understand its comparative worth except those faced with a scarcity. When water becomes scarce, use becomes efficient. Although some countries now partially accept the concept of regional management of the water supply, international and continental management remains novel. Yet this approach must be taken if the water needs of the next two hundred years, and for a population of fifteen billion, are to be met. The time scale for effective investment requires long-term intergenerational statesmanship and commitment. The problem is not the availability of technology or water.

The quantitative aspects of the water supply have been widely studied and, although complex for specific regions, can be viewed more simply on a global basis. Nature annually contributes in fresh water to the world's continents about ten times the world's water use for all purposes. On average, about 70 percent flows to the sea in seasonal floods; even if some could be captured, its storage might be an insuperable problem. Of the 30 percent that is theoretically usable, only about two-thirds is reachable from inhabited areas, so our potential global resource is, therefore, 20 percent of nature's original provision. Double the present world usage, will this amount suffice when the global population increases three-fold? Obviously, in the next two centuries the efficient use of water, its management, and the technology of recycling and desalination must all take part in achieving an adequate water supply.

The present use of the world's water breaks down as follows: approximately 63 percent for agriculture, 25 percent for industry, 7 percent for cities, and about 5 percent for system losses. Drinking demands a very minor share of water. In the United States the present distribution of the managed water supply works out to about 47 percent for power-plant cooling, 34 percent for irrigation, 10 percent for municipal and domestic use, 7 percent for manufacturing, and 2 percent for losses. Of these US uses, about one-fourth disappears through evaporation, transpiration, and product inclusion, leaving about three-fourths that is theoretically recoverable for recycling. The bulk of the cooling water in power plants already returns to rivers and lakes. So, in the United States, about 17 percent remains available for recycling from municipalities and industries.

Agriculture not only uses most of the world's water, it is also the most profligate. Crops actually take up only about a third of the water used for irrigation; the rest is lost. Microirrigation techniques can now reduce most of these losses. In the United States this would release about 20 percent of the managed water supply for other uses. Industrial and municipal use can be significantly reduced by recycling wastewater. That is feasible, however, only if the cost of removing the pollutants is not too high. Within manufacturing processes, such internal recycling may be relatively inexpensive. Strategically and economically, agricultural and fertilizer runoff, human waste, and industrial effluent offer the best opportunity to minimize pollution. Dealing with these at the sources rather than at the points of reuse lessens costs.

With careful management, investment in technological aids, and political skill, the water needs of a global population of fifteen billion could be met. Water already stresses many localities today, and action in these areas could demonstrate possible strategies. If water is valued highly enough, several conservation and supply options arise, such as more efficient water use in agriculture. Conventional aqueducts can bring water to people in need, or, conversely, people can migrate to water sources. Capturing and storing river runoff has a long history and certainly can be expanded, but not without some ecological cost. Constructing water infrastructures and changing water policies can take decades—raising again the problem of response times mentioned at the outset of this essay.

The ocean offers a final abundant source, and desalination will always be a technical option for fresh water if energy is available. For numerous US seaside communities, the cost today of desalination often roughly matches that of pumped aqueduct water, but the electricity demand is large for both—about a 25 percent increase in the average electricity load for a seaside city. All the technical options for increasing the water supply to the consumer require substantial energy inputs, mostly in the form of electricity. The next centuries will intimately link long-range water and energy strategies.

Energy

In their search for economic growth, today's less-developed countries (LDCs) will largely determine the future global energy use. Per capita energy growth is strongly associated with economic growth. Per capita, LDC residents now use one-fifteenth of the US level, while the global average per capita of all countries is one-fifth of the current US level. In two hundred years the global average might reach today's US level, with some countries naturally still lagging well behind the average. This growth works out to a rate of 1.4 percent per year, a modest expectation for global economic growth. Threefold population and five-fold per capita energy increases would together multiply global energy demand 15 times. Even if efficiency improvements cut in half this level of energy demand, a 7.5-times increase still boggles the mind.

However, in the past two hundred years, annual global energy use has increased approximately 50 times, primarily through the use of mineral fuels. From this perspective, an increase of 7.5 times in the next two hundred years may be a modest projection and challenge. We already foresee with some confidence more than a 2.5-fold multiplication in global energy demand from the present level by the year 2060—unless global economic growth is substantially suppressed. The multiple is less significant than its implications. The notion promoted by some environmentalists that global energy demand can be stabilized indefinitely at present levels is manifestly unrealistic, unless they concomitantly condemn most of the world's people to perpetual poverty.

In two hundred years all fossil fuels will still be available but more costly, as less-accessible geologic reserves are developed. Recall that global energy consumption fifty years ago was about one-fifth of today's consumption, and the perception of economically recoverable resources, especially for oil, was also much more limited than today's. Many experts of that era predicted a serious scarcity of low-cost fuel during the 1960–1975 period. The history of natural gas and oil predictions in the United States illustrates the fallibility of current estimates of reserves and resources. During World War I, the US government issued public notices that the demand for natural gas was exceeding the supply and warned that a return to expensive manufactured gas loomed. In 1920 the US Geological Survey predicted that all oil reserves would be depleted in fourteen years. By 1960, for every barrel believed available in 1920, eight had been extracted and five more had been proven to exist. In 1974 the Federal Power Commission stated categorically that the United States was reaching down to the "dregs" of its natural gas resources with "drastic" and "momentous" implications. This incorrect assessment formed one basis for the Carter administration's decision to restrict gas as a fuel for electricity generation, an action reversed a decade later in view of the obvious and continuing abundance of gas. In fact, gas is now often the first choice for electricity generation, economically and environmentally.

Fuel resources are not a static quantity. The technology of resource exploration, development, and extraction continuously improves in response to economic incentives, thus avoiding the scarcity. The cost of oil and gas today is about the same as fifty years ago, and proven reserves have doubled in spite of massive interim production. Nevertheless, today's known oil and gas reserves appear inadequate to meet all global energy demand for the next two hundred years. Based on atmospheric oxygen as the by-product of eons of photosynthesis, Earth's fossil-fuel content, including coal, is estimated at roughly 30,000 times the world's present annual rate of consumption. Technology will make some of this additionally available in the next two centuries, by such measures as opening the resources of the deep-oceans' bottoms.

Biomass fuels and hydropower, at their feasible limits, might provide as much as 20 percent of global energy demand by the year 2200 (i.e., equal to about 1.5 times today's global total energy demand) provided that they do not seriously compete for land with food and fiber production. The intermittent renewables, solar and wind, might provide another 20 percent, although they remain expensive, awaiting breakthroughs in the cost of energy storage. Growth in fossil-fuel and nuclear sources will supply the balance. Nuclear power (fission or fusion) may be the only source feasible for large-scale expansion because of its relative environmental cleanliness, in spite of the current public fear about its novel safety issues.

Of course, in two hundred years we may discover some now-unknown source, or tap "hot rocks" from the Earth's mantle, or focus solar space mirrors on Earth collectors, or invent and build a new energy storage or global superconducting grid that makes solar power more important. Such speculations titillate but cannot form the basis for responsible long-range strategies today. In the last thousand years the only new primary energy source discovered was nuclear power. We should not expect that the next two centuries will provide another such discovery for use in a future "post-fossil fuel era," although we know that we cannot foresee many of the potentials of technology.

As the economic development of the world continues, the fraction of primary energy used to make electricity as an intermediate will increase (see Ausubel and Marchetti, this volume). Electricity is the lifeblood of economic growth and technologic development in modern societies. More than a third of today's global primary energy input goes to electricity generation; in the next half century it is likely to rise to more than half, even with improved energy-to-electricity conversion efficiencies. New technologies create new electricity demands. For example, information systems, computers, and televisions now consume a sizable fraction of the electricity in industrial countries. Large-scale water desalination and purification, large-scale recycling and reprocessing of materials, and nonpolluting transport are already visible areas for growth in electricity demand. Electrification will dominate global energy systems in two centuries.

This energy scenario emphasizes our present social responsibility to maintain viable roles for every feasible future energy option, especially nuclear power. No single fuel will satisfy needs for the next two centuries, and promoting a single path is irresponsible and grossly deceptive to the public. For example, to go beyond the generous solar and wind estimate given above, the energy storage costs (and their inefficiencies) would drain the world's capital resources. The end of superpower confrontation has outdated a strident antinuclear posture, primarily a by-product of historic antipathy to nuclear weapons and secondarily a gesture against a novel and rapidly growing technology whose risks needed to be better understood. With the prospect of long-term growth in energy demand, such ideological burdens should be removed from responsible long-term strategic perceptions and from the real problems of implementation facing all supply options. Societies inexorably need energy, particularly electricity. The challenge in the next two centuries is to supply this while meeting the constraints of economics, politics, environment, health, and safety.

Human Settlements

An acceptable quality of life encompasses much more than a bowl of food and a cup of water. We also care where we dwell. Can the world make room for fifteen billion people? Land space is obviously ample if one considers the huge unoccupied areas of the world, excluding the polar ice caps. However, these areas are empty because people consider them uninhabitable, usually due to a lack of water or to climate extremes. At some cost, as discussed earlier, we can mitigate both of these limitations. Whether societies will make investments to open new lands for major settlements remains uncertain. Water converted large areas of California from semidesert to farms and cities. This sort of land expansion could surely continue to accommodate agriculture or popular climatic and scenic preferences.

The greatest shift in settlement patterns continues to be urbanization. As agriculture industrializes, the fraction of a country's population working in the fields decreases from as much as 85 percent to 3 percent or so. The resulting migration has created a strategic dilemma. Almost 50 percent of the world's people live in urban centers. About 60 percent of the North American population lives in cities of more than a million people. On the one hand, ugly, land-eating "urban sprawl" describes the resulting enlargement of many of the cities. On the other hand, the negative effects of crowding on the urban quality of life also show in the high-density center cities. The dilemmas will worsen with more population growth unless new trends in city life-style and design develop.

Population density, a crucial factor, varies greatly among cities. Most of the cities in the United States and Europe are of similar densities, probably because of their similar architectural concepts: New York is typical, with about eleven thousand people per square mile. Population density in Hong Kong is 21 times greater than New York; Jakarta and Bombay 11 times; Cairo 9 times; Mexico City 3.5 times; and Tokyo only 2 times, although with thirty million people (about twice the number of metropolitan New York), it is the largest city in the world. These cities are growing at annual rates of from 1 to 3 percent, mostly by expansion at the fringes. If present trends continue, about 80 percent of the world's threefold population increase during the next centuries will be concentrated in large cities—a fivefold increase of urban population. Will this raise density or "sprawl"?

Two opposite strategies become evident in response to the problem. The first is technology-based centralization; the second is counterurbanization. Where land is precious, as in Japan, the first is being pursued through many super-high-rise buildings. The population of Tokyo equals that of California, but it is crowded into less than 1 percent of California's area. As another measure, Tokyo has 2.7 times the population of Los Angeles in about the same area. The Japanese seek relief by constructing super-high-rise buildings with integral urban services, so that they work as minicities. One extreme proposal is for a five-hundred-story high rise (5 times the height of the World Trade Center in New York) that could house more than one hundred thousand people. Japan already has both ski slopes and bathing beaches enclosed by huge domed structures (see Ausubel, this volume).

In many big cities, high-rise apartment and office buildings are replacing low-level homes and shops. They also pose potentially costly civic burdens. Because of their high population density the buildings are vulnerable to the catastrophic effects of major equipment failures, and they create complex demands for traffic, policing, fire, and other services for the city of which they are a part. The response to the terrorist bombing of the World Trade Center in New York exemplifies these problems. The historic motivation for continued urbanization arises from the benefits of juxtaposing important facilities and services, such as large markets, hospitals and health services, theaters, educational institutions, and administrative services. The time value of travel for these services (their convenience) seems most important.

More recently, technology is stimulating the second option— counterurbanization. Telecommunication, both verbal and visual, is becoming simpler and more prevalent, so that any service that does not require physical contact can be made available at any distance with no delay and at modest cost. High-speed roads, rails, and airways can shorten travel times for physical exchanges. The technology of transport is the key to counterurbanization. For example, networks of high-speed electric trains could spread development over regions rather than concentrate it in one city. Trains dominated in the early 1900s but were overcome by the popularity of the automobile, which provided faster door-to-door travel. National resources were invested in roads rather than in rails. The technology of high-speed rail systems, and perhaps improved helicopter or vertical takeoff flight, will make these important modes of travel in the coming centuries.

In the absence of fast means of penetrating the center city, the trend will grow for subcities at the periphery of large centers, often ranging into the rural areas. A better quality of life at a lower economic cost has made these increasingly attractive relative to urban living. As numerous subcities develop around adjacent urban centers, they overlap into a regional megalopolis of relatively low population density, as in the northeastern United States. This counterurbanization has been further stimulated by the increasing urban disamenities of social and environmental origin, which detract from the historically high productivity of urban living.

Roads will never become obsolete, and they can also encourage counterurbanization. Perimeter loops around the big urban centers have had startling effects in the past several decades. Wherever these perimeter loops intersect a radial road from the center, new communities, businesses, and manufacturing have developed. These areas then become the centers of smaller clusters. The most recent urban road systems include both an inner loop and an outer loop, as well as intersecting radials. Apparently a road system that minimizes travel time from outlying communities to urban facilities encourages decentralization of the population. The spread provides the several amenities common to suburban living. From a national resource perspective, the requisite travel infrastructure may be a desirable investment. But it fosters a continued reliance on the personal automobile rather than on public transport. Is this pattern desirable for the next two centuries from the viewpoints of energy resources and the environment? Perhaps the electric vehicle and nuclear power can resolve the question.

The influence of technologic changes in shaping human settlements is clear. Looking forward two centuries, the threefold increase in global population poses the question of which strategies and patterns should be encouraged. Both urban minicities and counterurbanization subcities will continue to have available increasing technologic options and capabilities. Is it wise to support them? Should we encourage a life-style predominantly under covered domes, in a high-tech environment, shielded from nature's extremes? Should this trend be accepted as inevitable?

Both of the long-range regional strategies will undoubtedly be pursued. Both require large-scale planning and investment. A cynic might comment that a democratic society is rarely capable of deciding between such alternatives and that both will be followed until one haphazardly becomes the choice. A practical observer would note that the urban choices of the past endure for a very long time, as evidenced by the remnants of early transport systems, and that the free competition of alternative developments does not ever exist. With diverse motives, governments have always intervened to support one of the prevailing concepts. The world's major cities show the persistent remnants of errant favors. In light of the worldwide trend for people to drift to urban centers, the forms that such centers should eventually seek warrants much consideration. In these matters strategies are made, either explicitly or implicitly, by the ongoing decisions of governing bodies about budgets, regulations, zoning, and standards, as well as by the emergence of technical possibilities.

OVERVIEW

This attempt to foresee the major constraints on global societies in the next two hundred years and how technologies may lift them touches on only the most obvious. On a global basis, natural resources do not appear as barriers to a habitable and sustainable world. However, social, political, and economic obstacles abound. Some regional management will always be needed to ensure clean air and water, unpolluted food, and physical security. Technologies with which to meet these needs exist today. The shared nature of the needs clearly requires government leadership to protect the common interest.

The systems providing food, water, energy, and habitat depend on a mix of choices by individuals, enterprises, and government. The final acceptance or rejection of goods and services is made by individuals. The options are limited by the outcomes of government constraints that are placed upon the development of alternatives. These long-term choices epitomize the philosophical dichotomy between those who believe in the efficacy of government planning by "command and control" and those who believe that "free markets," with minimal government constraints, will result in optimal economic and technical development. Once a government intervenes to set constraints, presumably in the public interest, the field for free competition is no longer level, if it ever was, and some options will be inequitably treated.

However, ample history of damage to the common resources of all, such as air, water, and environment, indicates that truly unconstrained free competition tends to focus on short-term values and neglects the long term. Moreover, we know that even with the best of intentions most decisionmakers, public and private, are unable to make choices today that are likely to be adequate for a highly uncertain future. The balancing of minimal government intervention with some cautionary constraints on free economic competition is the obvious goal. Achieving that balance is a complex political process with uncertain outcomes, particularly in the democratic political systems that we all cherish. Achieving the ideal of government by an informed public is hard when the information base for choices is usually uncertain, and winners and losers are involved.

I have not dealt with the substantial environmental impacts of the growing world population and economy, seen by many as major threats to both ecosystems and human health. They seem to me within the capability of technology, today's and tomorrow's, to manage during the next two hundred years and therefore matter less for our intergenerational accountability. Regarding present concerns that earn global rather than regional attention, I do not anticipate that issues such as climate change, ozone depletion, and loss of biodiversity will cause life-threatening crises. We should study, anticipate, and respond to them, as we do with global diseases. I expect that a succeeding global generation, with its basic survival needs met and with superior knowledge and resources to implement technical options, will respond to such environmental issues by mitigation and adaptation—if its own priorities at that time motivate it to do so. I have little faith in the wisdom of governments to choose correct long-term strategies, particularly in light of the unforeseen stream of scientific and technical changes that usually make preventive interventions for such problems unsound and damaging. I do not believe in manipulating the options for future generations—it presumes a certitude of foresight that no one has. Clearly this is a controversial view challenged by those fearful of environmental outcomes, uncertain as they may be.

A scenario always represents the author's views, and this one is no exception. I am an optimist about technology's ability to maintain the global availability of the resources needed to sustain a large population. I am a pessimist about the ability of elite "brain trusts," scientific or political, to plan globally much beyond the present. I am very pessimistic about the ability of world governments to formulate global plans centrally when they are deeply deficient in managing their own domestic issues. So, I urge minimal governmental interference in the management of global resources and maximum freedom for the development and use of technical options.

The real threats to a habitable and sustainable world in the next two centuries arise from the continuing social turmoil associated with the relatively inflexible cultural and ethnic differences among people. The seventy-year campaign of the USSR to wipe out such differences apparently failed completely. If the past two centuries are indicative, the magic cure for self-destructive social conflict has not been found. The technologist looks for clean water and air, sanitary sewers, and security from nature's blows. All these contributions to the habitability of the planet are quickly negated by the misery of wars and social conflicts. Universal economic prosperity is often suggested as the simple cure. It would certainly help, but is it sufficient? It seems that an optimistic sociologist and an optimistic economist are necessary to bolster my technological optimism for a globally happy future.

REFERENCE

Walter, D., ed. 1992. Today Then: America's Best Minds Look 100 Years into the Future on the Occasion of the 1893 World's Columbian Exposition. Helena, Mont.: American & World Geographic Publishing.

Biographical Data

JESSE H. AUSUBEL is director of the Program for the Human Environment at The Rockefeller University in New York City. Mr. Ausubel's interests include environmental science and technology and industrial evolution. The main themes of the Rockefeller program are industrial ecology (a field Mr. Ausubel helped originate in 1988–1989) and the long-term interactions of technology and the environment. From 1977 to 1988, Mr. Ausubel was associated with the National Academy complex in Washington, D.C., first as a resident fellow of the National Academy of Sciences, then as a staff officer with the National Research Council Board on Atmospheric Sciences and Climate, and from 1983 to 1988 as Director of Programs for the National Academy of Engineering. Educated at Harvard and Columbia universities, Mr. Ausubel serves concurrently as a program officer of the Alfred P. Sloan Foundation, where his concerns include the performance of the U.S. academic enterprise.

ROBERT A. FROSCH is a senior research fellow at the Center for Science and International Affairs of the John F. Kennedy School of Government at Harvard University. In 1989 Dr. Frosch revived, redefined, and popularized the term "industrial ecology," and his research has focused on this field in recent years, especially in metals-handling industries. In 1993 he retired from the position of vice president of the General Motors Corporation in charge of the North American Operations Research and Development Center. Dr. Frosch holds a Ph.D. in theoretical physics from Columbia University. After doing research in underwater sound and ocean acoustics, he served for a dozen years in a number of government positions, including deputy director of the Advanced Research

Projects Agency of the Department of Defense, assistant secretary of the Navy, assistant executive director of the United Nations Environment Program, and administrator of the National Aeronautics and Space Administration. He is a member of the National Academy of Engineering.

SHEKHAR GOVIND is assistant professor of civil and environmental engineering at the University of Texas at Arlington, where he teaches transportation planning, urban infrastructure, and systems analysis. Dr. Govind received his bachelor of technology degree in civil engineering at the Indian Institute of Technology, Delhi, and his master's and Ph.D. degrees in civil engineering from the University of Texas at Austin. His research interests include intermodal transportation systems, urban planning and growth, and design of algorithms for parallel processing computers.

ARNULF GRÜBLER is a research scholar in the Environmentally Compatible Energy Strategies Project at the International Institute for Applied Systems Analysis in Laxenburg, Austria. Dr. Grübler's scientific interests include the long-term qualitative and quantitative history of technological evolution and its relationship to global change, as well as the dynamic modeling and analysis of technical change within and between different economic and social environments. He has built up a database of more than a thousand cases of diffusion of technologies, especially in energy and transportation. Dr. Grübler received his master's degree in engineering from the Technical University of Vienna, where he was also awarded his Ph.D. He lectures regularly at the University of Vienna, the Technical University of Graz, and the International Centre for Theoretical Physics, Trieste.

ROBERT HERMAN is L. P. Gilvin Centennial Professor, Emeritus, in Civil Engineering and sometime professor of physics at the University of Texas at Austin. Before assuming his present position in 1979, Dr. Herman headed the Department of Theoretical Physics and the Traffic Science Department of the General Motors Research Laboratories. His research has covered a wide range of both theoretical and experimental investigations, including molecular and solid-state physics, high-energy electron scattering, astrophysics and cosmology, and operations research, especially vehicular traffic science and transportation. He led a National Academy of Engineering project on infrastructures that resulted in the 1988 National Academy Press book *Cities and Their Vital Systems* (J. H. Ausubel and R. Herman, eds.). With Ralph Alpher in 1948, Dr. Herman made the first theoretical prediction that the universe should now be filled with cosmic microwave background radiation, which is key evidence for the validity of the big bang model of the origin of the universe. He holds degrees in physics from

City College, New York, and from Princeton University and is a member of the National Academy of Engineering.

ROBERT W. KATES is an independent scholar in Trenton, Maine, and director emeritus of the Feinstein World Hunger Program at Brown University. His interests include the prevalence and persistence of hunger, long-term population dynamics, sustainability of the biosphere, and natural and technological hazards. Professor Kates worked in a steel mill in Gary, Indiana, and received a Ph.D. degree from the University of Chicago without having received an undergraduate degree. From 1962 to 1986 he served on the faculty at Clark University. From 1967 to 1969 Professor Kates directed the Bureau of Land Use Planning in Dar es Salaam, Tanzania. He is a member of the National Academy of Sciences and a recipient of the Presidential Medal of Science.

CESARE MARCHETTI is an Institute Scholar at the International Institute for Applied Systems Analysis, which he joined 1974 to contribute to research on energy systems. Dr. Marchetti received his education in physics at the University of Pisa. His early work was in physical chemistry and chemical engineering of nuclear reactors. He joined the European Atomic Energy Community (EURATOM) in 1959 and represented EURATOM in Canada for 2 years, working in the field of reactor optimization. He returned in 1961 to head the European Community Ispra Research Center Physical Chemistry Division, and later the Materials Division, where his group pioneered the exploration of the production of hydrogen by thermochemical and other means. Since the early 1970s he has authored more than 100 papers on the time patterns of technological choice, their conceptual origins, and their implications for future developments in energy, transportation, and the spatial organization of human activities.

KLAUS MICHAEL MEYER-ABICH is a professor of the philosophy of nature at the Wissenschaftszentrum (science center) in Nordrhein-Westfalen and at the University of Essen. He studied physics and philosophy at Hamburg and Gottingen in Germany and at Bloomington, Indiana, and Berkeley, California, in the United States. He received his M.S. degree in physics in 1961 and his Ph.D. in 1964. Dr. Meyer-Abich has held faculty positions at the University of Hamburg and at the Max Planck Institute for the Study of the Conditions of Life in the Modern World at Stamberg, near Munich. He has also served as a minister for science and research in the state of Hamburg and as a member of the German parliament's Enquête Commissions on the Future of Energy Policy and on the Protection of the Atmosphere.

NEBOJŠA NAKIĆENOVIĆ directs the Environmentally Compatible Energy Strategies Project at the International Institute for Applied Systems Analysis in Laxenburg, Austria. He also leads energy scenario development for the World

Energy Council and has been a principal contributor to the energy studies of the Intergovernmental Panel on Climate Change. His research interests include the representation of technical change in economic modeling, the long-term patterns of technological change and economic development, and, in particular, the evolution of energy, automotive, and aerospace technologies. Dr. Nakićenović holds bachelor's and master's degrees in economics and computer science from Princeton University and the University of Vienna, where he also received his Ph.D.

LEE SCHIPPER is a visiting scientist at the International Energy Agency, on leave from the Lawrence Berkeley Laboratory, University of California, Berkeley, where he has been a leader of the International Energy Studies group. He has consulted for the United Nations, the World Bank, and many other international organizations and has worked in or visited more than 30 countries in connection with his work. His current work includes comparisons of energy use in the manufacturing and transportation sectors of major industrialized countries. Dr. Schipper holds degrees in music and physics from the University of California, Berkeley. He is a jazz vibraphonist and pianist and has been a faculty member of the San Francisco Conservatory of Music.

CHAUNCEY STARR was founding president and vice chairman of the Electric Power Research Institute, where he is now president emeritus. Following a 20-year career in industry, during which he was vice president of Rockwell International and president of its Atomics International Division, Dr. Starr served as dean of the School of Engineering and Applied Science at the University of California, Los Angeles. He made pioneering contributions to nuclear propulsion for rockets and ramjets, miniaturization of nuclear reactors for space, and development of nuclear power plants. Dr. Starr received a degree in electrical engineering and a Ph.D. in physics from Rensselaer Polytechnic Institute. He is a member of the National Academy of Engineering and in 1990 received the National Medal of Technology for his contributions to engineering and the electric industry.

PAUL E. WAGGONER is a Distinguished Scientist at the Connecticut Agricultural Experiment Station in New Haven. He has held that title since 1987, after serving 15 years as director of the station. Dr. Waggoner was educated in meteorology and was a weather forecaster in the U.S. Air Force. He studied agricultural climatology and plant pathology at Iowa State University, where he received his Ph.D. His investigations have encompassed water relations and diseases of plants as well as micrometeorology. Since 1983 he has participated in studies of climate change with the National Research Council, the American Association for the Advancement of Science, and the Council for Agricultural Science and Technology. He is a member of the National Academy of Sciences.

IDDO K. WERNICK works as a research associate in the Program for the Human Environment at The Rockefeller University. Dr. Wernick received his bachelor's degree in physics from the University of California, Los Angeles, and his Ph.D. in applied physics from Columbia University. Dr. Wernick's research covers the environmental consequences of natural resource use and environmental analysis of materials flows in the U.S. economy. He is currently investigating materials flows as they relate to industrial practice in the forest products sector. His recent work includes preparation of a research agenda for the emerging discipline of industrial ecology. Dr. Wernick has also written on the technical and political context for community assessment of local environmental risk.

ROBERT M. WHITE is a senior fellow of the University Corporation for Atmospheric Research, president of the Washington Advisory Group, a Washington, D.C., consulting firm, and president emeritus of the National Academy of Engineering. Dr. White established one of the first private corporations devoted to environmental science and services. He served in the federal government under five presidents, from 1963 to 1977, first as chief of the U.S. Weather Bureau and finally as the first administrator of the National Oceanic and Atmospheric Administration. In these capacities he is credited with bringing about a revolution in the U.S. weather warning system using satellite and computer technology, helping to initiate new approaches to the balanced management of the country's coastal zones, and promoting the protection of American fisheries. Dr. White holds a bachelor's degree in geology from Harvard University and a master's and Ph.D. in meteorology from the Massachusetts Institute of Technology.

Index